The 8th International Conference on Fluid Sealing was sponsored
and organised by BHRA Fluid Engineering and was held at the
University of Durham, England from 11th–13th September, 1978.

Acknowledgements

The valuable assistance of the panel of referees and the
Organising Committee is gratefully acknowledged.

Organising Committee

Dr. E.T. Jagger (Chairman)
R.A. Grimston, Flexibox Ltd.
B.D. Halligan, James Walker & Co. Ltd.
Dr. B.S.Nau, BHRA Fluid Engineering
Professor J.P.O'Donoghue, Northern Ireland Polytechnic
H.S.Stephens, BHRA Fluid Engineering
J.A.Stephens, Bestobell Seals Ltd.
N.G. Guy, BHRA Fluid Engineering

CONTENTS

OPENING ADDRESS

Professor F.T. Barwell
University College of Swansea

That this is the eighth conference on fluid sealing is indicative of the maturity of the subject and today's wide international representation is evidence of a healthy scientific status. However, when regarded from the standpoint of history, the science of fluid sealing was very late in developing and it is still largely the province of specialists rather than part of the stock in trade of mechanical engineers generally.

Bearings have received much more attention than seals, partly because of the discoveries of Beauchamp Jones (placed on a sound theoretical basis by Osbourne Reynolds) and partly because the consequences of a failed bearing are usually more spectacular than those of a leaky seal. Indeed, although I had worked for some time on bearings (now Tribology) I only became aware of the scientific approach to sealing with the work of White and Denny at Imperial College. Due to the foresight of Lionel Prosser, the first director, this was continued in the newly formed British Hydromechanics Research Association which has, in consequence, attracted a world-wide leadership in the field.

Most of us learned about seals, glands and gaskets from our mates when we were apprentices. We learned to scrape metal-to-metal joints to withstand great pressures, to pack glands with hemp soaked in tallow, with asbestos-based proprietary materials or mysterious patented compounds. This practical training was effective and it was a matter of pride that a steam locomotive for example should leave the sheds without a whisper of steam escaping from the piston or valve rods, even at the highest speeds. There was, however, at least as far as I am aware, no corresponding theoretical instruction. There were, in the drawing class it is true, models of successful stuffing boxes which we learned to copy, but no theory. Although it was apparent the surface form and finish were vitally important, we then had no objective measure of what was considered smooth.

Nowadays we can class dynamic seals as tribological systems and much of the theory which has been developed for bearings, such as elasto-hydrodynamics, can be applied to them. We are fortunate that Professor Higginson, a pioneer in this field, will be joining us for dinner tomorrow.

This is also the case with operational techniques, notably the condition monitoring of a machine by studying the metallic particles present in the lubricating oil. In one method, known as "Ferrography," particles are precipitated in a magnetic field so that they may be analysed both quantitatively and qualitatively to denote the occurrence of objectionable wear, before serious failure can result. This is important economically because it can avoid much expensive "preventive maintenance" by allowing machines to operate for extended periods in the knowledge that all the bearing systems are operating normally. In hydraulic machinery, safe operation often depends upon the integrity of polymeric seals as much as upon interacting metal surfaces. However, being non-magnetic, particles arising from their degradation escape detection by the Ferrographic techniques. An exciting recent development employs means of treating solutions of the hydraulic fluid samples so that the polymer particles are made para-magnetic and can be detected by existing methods. Thus the incipient feature of a polymeric seal can be detected from wear debris in samples of the hydraulic fluid and remedial action taken. Conversely, if the plant can be given a clean bill-of-health expensive inspection by stripping can be avoided.

I instanced this to demonstrate that we are concerned with a rapidly developing technology from which new concepts are continuously being born. I am thus confident that your three days of discussion will be lively and the interaction of so many who have made fluid sealing their speciality will be particularly fruitful in new concepts. I therefore open your conference with every confidence of a successful outcome. The conference is now open.

LIST OF PAPERS PRESENTED

The majority of the papers are published in Volume 1. Those marked * are to be found in this Volume.

INVESTIGATIONS CONCERNING RADIAL LOAD OF RADIAL SHAFT LIP SEALS

F. Ridderskamp

Carl Freudenburg Simrit-Werk
Federal Republic of Germany

Held at the University of Durham, England.

Conference sponsored and organised by BHRA Fluid Engineering

©BHRA Fluid Engineering, Cranfield, Bedford MK43 0AJ, England.

1. INTRODUCTION

Radial seals of synthetic rubber have been used in the automotive industry and general construction for sealing media (mineral oils and greases) for about 40 years.

With the engineering progress, mainly in the motor car sector after 1945, the demand for improved radial seals increased.

The "standard shaft seal" of NBR rubber, which for a long time was predominant at all sealing areas, did not meet all the increased requirements any more. After thermically and chemically resistant rubber compounds had been developed for sealing at high temperatures with partly alloyed oils (additive packages) and due to the increased attention paid to the profile design of radial seals, various methods and criteria were developed to help the research engineer in the rubber processing industry in designing radial seals.

2. AIDS FOR RESEARCH ENGINEERS IN THE SEALING INDUSTRY FOR AN OPTIMUM PROFILE

The profile of a radial seal is designed according to the following criteria:
2.1. Field experience – and test runs.
2.2. Results from ageing elongated standard elastomer specimen in various media.
2.3. Evaluation of the change in radial load (l) of radial seals after running tests.
2.4. Results of the change in radial load (l) after immersing radial seals in various media:
    a) extended
    b) not extended
    (Ageing in oil of the radial seal)
2.5. Change of radial load of a radial seal during operation.

3. OBJECT OF INVESTIGATION

3.1. Static and dynamic change in radial load of radial seals by ageing.

The object of this paper is to compare the change in radial load by using previous "static methods" (see 2.3 and 2.4) with the change in radial load during operation (see 2.5). Compared with the static methods the following factors were taken into account for the "dynamic methods":
    a) temperature field in the sealing lip
    b) tangential tension in the sealing lip
    c) wear of the sealing lip compound
    d) influence of mechanical ageing of the lip compound in the test rig (contact friction)

The object of the study was to replace the complicated "dynamic method" as regards the test device and measuring operation, by a more simple, equally effective measuring method.

3.2. To compare the change in radial load of various profile designs:
    a) profile type A (1966 ... 1974)
    b) profile type B (from 1974 on)
    and various compound variations of the three standard compounds:
    a) NBR
    b) ACM
    c) FPM

with each other, in order to receive further design criteria for the development of profile types for radial seals.

4. DESCRIPTION OF MEASURING METHOD AND TAKING MEASUREMENTS

4.1. Measuring method

In all test methods the radial load of the radial seals was determined by using the "Freudenberg radiameter" (2), shown in Figure 1.

The oil seal is pushed on a split mandrel with the nominal diameter of the running shaft. One half is fixed on the top of a stator, the other part on top of a leaf spring and thus causes a displacement of one half of the mandrel with respect to the other half fixed to the stator.

Consequently, measurement of radial load is reduced to the measurement of the deflection of the leaf spring. By proper dimensioning of the leaf spring the displacement of the two halves of the mandrel can be kept very small. The change of the diameter in relation to the nominal diameter of the shaft can be neglected. The displacement is measured with a dial test indicator, and the reading is taken about 2s after the seal is put on the radiameter.

4.2. Test Procedure

4.2.1. Changes in radial load for static immersion of radial seals in hot oil, not extended

Figure 2 shows how the measurements were carried out.
The radial seal was extended to the nominal diameter at various intervals and the radial load was determined.

4.2.2. Change in radial load for static immersion of extended radial seals in hot oil

Figure 3 shows how the measurements were carried out.

4.2.3. Change of radial load of a radial seal during operation ("dynamic method")

It was necessary to develop a special device to measure the radial load during operation at certain intervals. For the measuring operation the radial seal is removed from the device for a short period, is measured and then put back into the test device.

These measurements were carried out in a vertical design running rig with radiameter. Figure 4 shows the device employed for testing and measuring.

The measuring of the radial load takes place as follows:
    a) The test casing with the radial seal installed is lifted pneumatically so that the radial lip is pushed onto the material.
    b) The radial load value is taken from the dial test indicator of the radiameter.
    c) The test casing with the radial seal is pneumatically moved back to the test rig.

The reading of the measuring value takes about maximum 5 seconds. Thus it is guaranteed that the working conditions, above all the temperature, remain almost unchanged. Any possible measuring error is kept so small that it comes within the accuracy of reading the test indicator.

4.3. Test conditions

The conditions were adapted to the conditions customary in the motor car industry.
The **operating** time was:

$$t = 250 \text{ h}$$

The **temperatures**

$$\vartheta = 100^\circ \text{ for NBR compound}$$
$$\vartheta = 130^\circ \text{ for ACM and FPM compounds}$$

are usual working temperatures for radial seals in motor cars.
The **test oil**

$$\text{SAE 80 (MIL 2105)}$$

was an alloyed gear oil, mainly used in the gear-box of passenger cars and trucks.
The **speed** chosen (only for method 4.3) was

$$n = 3000 \text{ rpm}$$

This corresponds to the speed of a passenger car of about 100 km/h in fourth gear.

5. RESULTS OF COMPARATIVE INVESTIGATIONS WITH REGARD TO THE CHANGE OF RADIAL LOAD OF RADIAL SEALS

The measuring values obtained in the test series are shown in Fig. 5 - 13 and plotted as percentage change of radial load as a function of time.

The radial load value "100%" is the value of the new radial seal at room temperature (20°C). With this way of plotting the change of radial load in correlation to time becomes evident as a consequence of temperature, extension, influence of oil, number of rotations, wear of radial seal lip.

The material used were:

NBR1 a standard Nitrile used since 1968

NBR2 a special wear-resistant Nitrile

ACM 1 and FPM 1 are old standard ACM and FPM compounds used for all ACM and FPM applications in the period 1960 - 1974

ACM 2 and FPM 2 are new standard high wear resistance ACM and FPM compounds used for all ACM and FPM applications since 1974/5.

5.1. "Method 4.2.1." change in radial load with time of unextended radial seals statically immersed in oil

It is shown in Figs, 5 - 13 that radial load of the radial seals with spring and without spring decreases within the first hour of immersion to a value which then remains constant. A distinct difference can be seen in the decrease of radial load between a radial seal with spring and a seal without spring. The decrease of the radial load of the radial seal without spring is almost twice as large as the decrease of the radial load of a seal with spring. (Radial component of the spring is missing, thus "free" extension of radial seal lip possible.)

Results of measurements: Profile type A

	Compound	ΔPr (%) (Radial seal w. spring)	ΔPr (%) (Radial seal w/o spring)
Figs. 5 - 9	NBR 1 + 2	30	50
	ACM1 + 2	30 ... 40	65 ... 75
	FPM1 + 2	25 ... 40	50 ... 70

Results of measurements: Profile type B

	Compound	ΔPr (%) (Radial seal w. spring)	ΔPr (%) (Radial seal w/o spring)
Figs. 10 - 13	NBR 1 + 2	30	60
	ACM1 + 2	40	90
	FPM1 + 2	30 ... 40	40 ... 55

5.2. Change of radial load with time of extended radial seals during static immersion in oil (Method 4.2.2.)

Values for extended radial seals:
   a) Radial seals without spring $\approx 2.0\%$
   b) Radial seals with spring $\approx 3.5\%$

The change of radial load is very great during the first 30 hrs. approx. Thereafter the curves take an asymptotic course with a decreasing tendency (Figs. 5 - 13).

Results of measurements: Profile type A

	Compound	ΔPr (%) (Radial seal w. spring)	ΔPr (%) (Radial seal w/o spring)
Figs. 5 - 9	NBR 1 + 2	60	80 ... 90
	ACM1 + 2	50 ... 65	85
	FPM1 + 2	45 ... 55	60 ... 75

Results of measurements: Profile type B

Figs. 10 – 13	Compound	ΔPr (%) (Radial seal w. spring)	ΔPr (%) (Radial seal w/o spring)
	NBR 1 + 2	65	100
	ACM 1 + 2	50 ... 60	100
	FPM 1 + 2	50 ... 60	65 ... 85

5.3. Change of radial load during running test

Values for the extended radial seal (special test run/measuring device):
- a) Radial seal without spring ≈ 2.0%
- b) Radial seal with spring ≈ 3.5%

This method shows the "true" values of the change of radial load during operation with different profile types and elastomer compounds. Here, the curves showing decrease in radial load with time are flatter up to onset of the asymptotic region (after a test period of approx. 60 hrs.) than those found with method 4.2.2.

Results of measurements: Profile type B

Figs. 5 – 9	Compound	ΔPr (%) (Radial seal w. spring)	ΔPr (%) (Radial seal w/o spring)
	NBR 1 + 2	65	85
	ACM 1 + 2	50 ... 60	85
	FPM 1 + 2	45 ... 55	65 ... 80

Results of measurements: Profile type B

Figs. 10 – 13	Compound	ΔPr (%) (Radial seal w. spring)	ΔPr (%) (Radial seal w/o spring)
	NBR 1 + 2	65	85
	ACM 1 + 2	55	95
	FPM 1 + 2	50	70

5.4. Discussion of the test results:

5.4.1. "Static immersion of not extended radial seals in hot oil (Method 4.2.1.)"

This method is unsuited to evaluation of the change of the radial load during operation. Considerable differences become evident in the trend of curves and the values of the change of radial load compared with the methods below.

5.4.2. "Static immersion of extended radial seals in hot oil (Method 4.2.2.)"

This method is well suited to evaluate the change of radial load of radial seals during operation. There is only a little difference between the courses of curves, and the values of the changes of radial load differ only negligibly from the curves obtained with the "dynamic method".

5.4.3. "Dynamic method" (Method 4.2.3.)

The curves are almost identical in outline and value to those obtained through method 5.4.2.
The difference, the flatter curve of the decrease of radial load to the point of transition up to the asymptotic range, can be explained by the hardening of the seal face area due to excessive temperature at the seal edge and additional frictional heat during operation produced by the oil sump temperature (3). This increased temperature at the sealing area hardens the

surface, which delays the decrease of the radial load. This might also explain the fact that with the "dynamic method" the decrease of the radial load of the radial seal without spring in some cases does not amount to 100% as found during the investigation according to method 5.4.2. (NBR and ACM compounds).

6. Findings for the profile design of radial seals

Compared with profile type B, the profile type A (used in approx. 90% of all applications from 1966 to 1974) has a bigger rubber portion. The rubber therefore contributes a larger proportion of the radial load compared to the spring. The distribution of radial load between rubber and spring has been modified in the design of the profile type B (used in approx. 90% of all applications since 1974) so that the spring prevails over the rubber. This was achieved by the reduction in volume of the rubber profile.

Such reduction is of particular importance where compounds are exposed to a medium that produces swelling.

The test results for $\Delta Pr\%$ by methods 4.2.2. and 4.2.3. confirm this for NBR and very clearly for ACM.

Elastomer	Profile type/Seal type	Static method (4.2.2.)	Dynamic method (4.2.3.)
NBR	A/with spring	60	65
	A/w/o spring	85	85
	B/with spring	65	65
	B/w/o spring	100	85
ACM	A/with spring	60	55
	A/w/o spring	85	85
	B/with spring	55	55
	B/w/o spring	100	95
FKM	A/with spring	50	50
	A/w/o spring	70	70
	B/with spring	55	50
	B/w/o spring	75	70

7. Conclusions

7.1. Recommended method

Investigation has shown that method 4.2.2. (Static immersion, extended, in oil) and method 4.2.3. (Dynamic determination during operation produce almost identical values.

Consequently, it is advisable to use the simple and less costly method 4.2.2. for development work.

Factors for consideration in the 'dynamic method':
   a) Temperature field in the sealing lip
   b) Tangential tension in the sealing lip
   c) Wear of the sealing lip compound due to friction
   d) Influences of mechanically aged medium on the material of the sealing lip

Compared with the static method 4.2.2., (test period 250h, good lubrication of the sealing lip) these influences are not apparent in the range under investigation or only at the initial stage of the test. Using the dynamic method, the same decrease of the radial load is obtained with a delay of approx. 20 test hours, if the 30 hrs measured value is taken as the initial value.

The additional influences as per 7.1. a) - d) can be neglected considering the total result.

7.2. Radial load changes during operation

Radial seals of profile A and profile B operate under service conditions with a total radial load (rubber portion + spring portion) of approx. 40% after a period of about 50h (35% - 50%) compared with the initially radial load of the new radial seal at room temperature (ϑ room).

After approx. 50 working hours the rubber radial load portion drops to approx. 15% (0 – 30%) of the initial load.

7.3. Use in radial seal development

The results of the static and dynamic methods (4.2.2. and 4.2.3.) for the determination of the change of radial load during operation can be used for future designs of profiles and improvements and provide possibilities of comparison for new series of profiles.

8. References

1. Becker, B., Sealing of rotating shafts with radial oil seals, VDI-Z, Nos. 5 and 6, 1976, VDI-Verlag GmbH.

2. Schmitt, W. and Upper, G., Radial load as a lip seal design and quality control factor ASLE, Journal of Lubrication Tech., April 1968.

3. Upper, G., Temperature of Sealing Lips, Dissertation TH Karlsruhe 1968, Extract: "Automobil-Industrie", No. 1, 1970

Fig. 1   Freudenberg Radiameter

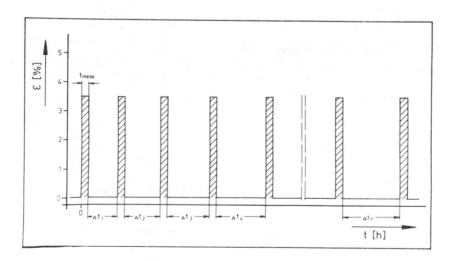

Fig. 2   Aging and Measuring of Shaft Oil Seals in Hot Oil.Method 1: Shaft Oil Seals without Permanent Extension

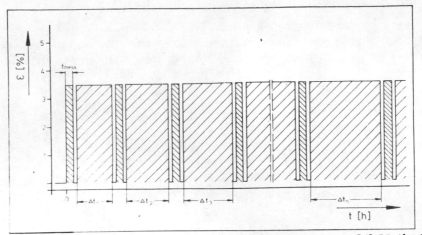

Fig. 3   Aging and Measuring of Shaft Oil Seals in Hot Oil. Method 2:
Shaft Oil Seals with Permanent Extension

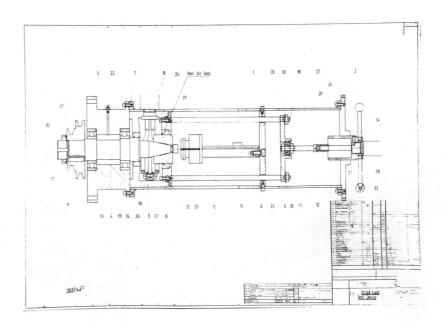

Fig. 4   Test Rig

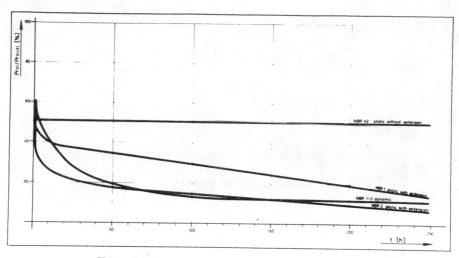

Fig. 5(a)
Profile - Type A Without Garter Spring
NBR 1 + 2 - 100°C, SAE 80

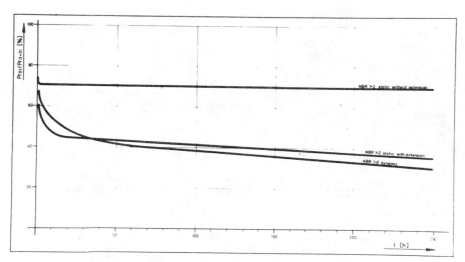

Fig. 5(b)
Profile - Type A With Garter Spring
NBR 1 + 2 - 100°C, SAE 80

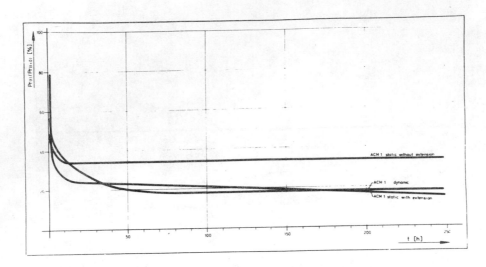

Fig. 6(a)
Profile – Type A Without Garter Spring
ACM 1 – 130°C, SAE 80

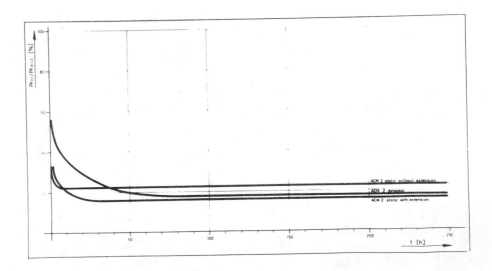

Fig. 6(b)
Profile – Type A Without Garter Spring
ACM 2 – 130°C, SAE 80

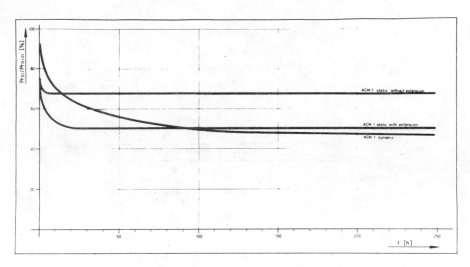

Fig. 7(a)
Profile – Type A With Garter Spring
ACM 1 – 130°C, SAE 80

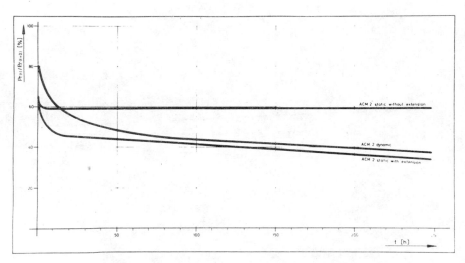

Fig. 7(b)
Profile – Type A With Garter Spring
ACM 2 – 130°C, SAE 80

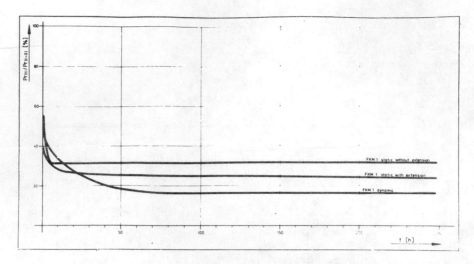

Fig. 8(a)
Profile - Type A Without Garter Spring
FPM 1 - 130°C, SAE 80

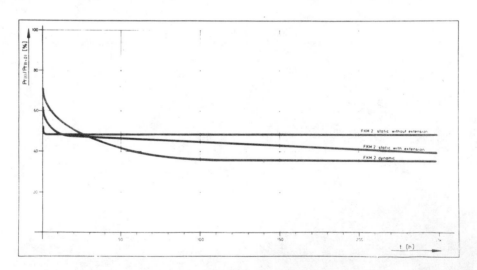

Fig. 8(b)
Profile - Type A Without Garter Spring
FPM 2 - 130°C, SAE 80

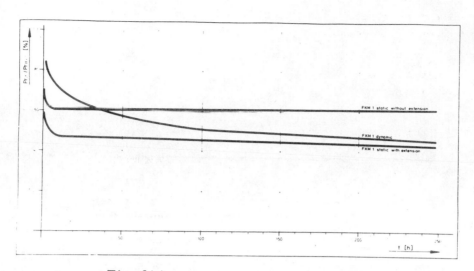

Fig. 9(a)
Profile – Type A With Garter Spring
FPM 1 – 130°C, SAE 80

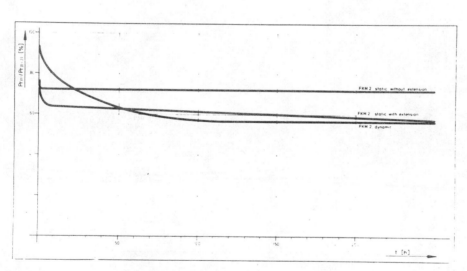

Fig. 9(b)
Profile – Type A With Garter Spring
FPM 2 – 130°C, SAE 80

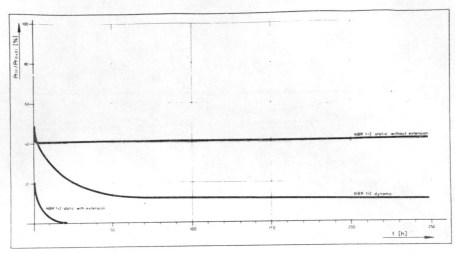

Fig. 10(a)
Profile - Type B Without Garter Spring
NBR 1 + 2 - 100°C, SAE 80

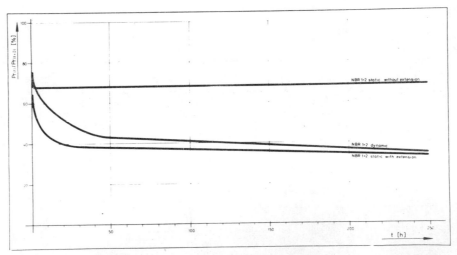

Fig. 10(b)
Profile - Type B With Garter Spring
NBR 1 + 2 - 100°C, SAE 80

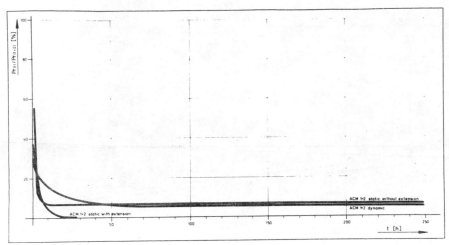

Fig. 11(a)
Profile - Type B Without Garter Spring
ACM 1 + 2 - 130°C, SAE 80

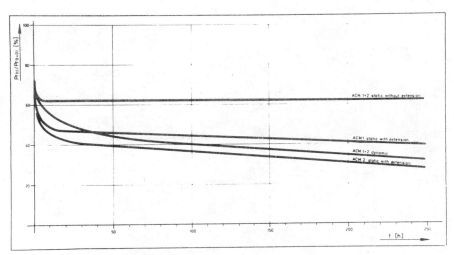

Fig. 11(b)
Profile - Type B With Garter Spring
ACM 1 + 2 - 130°C, SAE 80

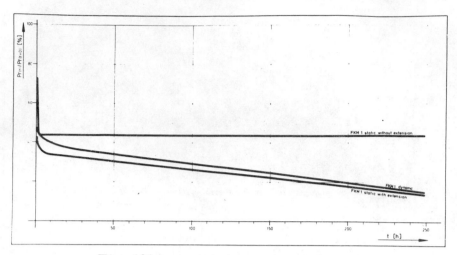

Fig. 12(a)
Profile – Type B Without Garter Spring
FPM 1 – 130°C, SAE 80

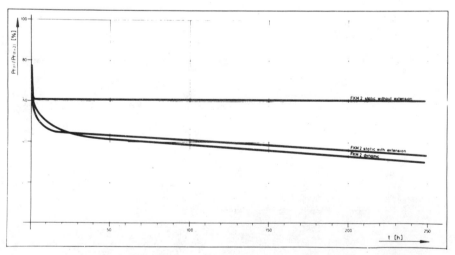

Fig. 12(b)
Profile – Type B Without Garter Spring
FPM 2 – 130°C, SAE 80

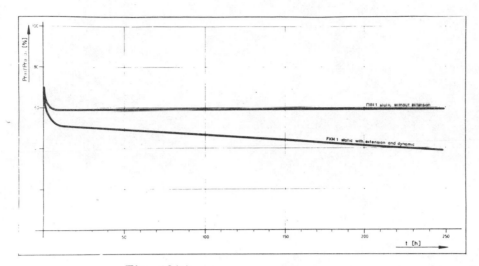

Fig. 13(a)
Profile - Type B With Garter Spring
FPM - 130°C, SAE 80

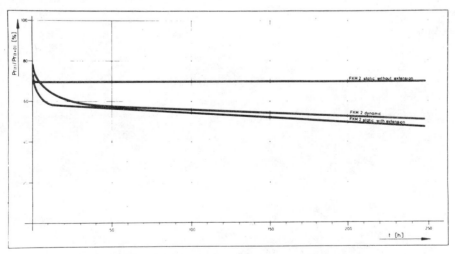

Fig. 13(b)
Profile - Type B With Garter Spring
FPM 2 - 130°C, SAE 80

MECHANICAL SEALS FOR AQUEOUS MEDIA SUBJECT TO HIGH PRESSURES

W. Schöpplein

Feodor Burgmann, Dichtungswerk GmbH & Co.
Federal Republic of Germany

Summary

Based on empirical values gained during practical use of heavy-duty seals employed for aqueous media, and backed up by Company development work and testbed results, the author gives a survey of design criteria applicable to high pressure seals, special consideration being given to problems of deformation.

When choosing a seal, precise acquisition of, and proper regard to, operational parameters is of decisive importance - disregarding certain marginal constraints programs for the failure of any seal. In addition to an analysis of said constraints, the author surveys the most common seal materials - steels, sliding face materials and elastomers-, giving technical data and showing some common forms of damage.

In conclusion, the author gives practical examples showing the design of high pressure seals to be used with aqueous media.

Held at the University of Durham, England.

Conference sponsored and organised by BHRA Fluid Engineering

©BHRA Fluid Engineering, Cranfield, Bedford MK43 0AJ, England.

I.   Introduction

Mechanical seals are by now well-proven and safe, low-mainte-
nance and cost-effective rotary-shaft seals. In connection with
optimization of manufacturing technology in process engineering,
with increasing generating capacity of conventional and nuclear
power plants, and with the evolution of new technologies such
as coal gasification, to name just a few examples, demands made
upon the sealing elements to be used keep increasing. Problems
encountered when operating and connecting multi-stage high-
pressure seals, resulting in customer demands for single-stage
pressure reduction whenever possible, underline a recent trend
towards increasing loads on individual seals.

According to the classification of seals made by E. Mayer /1/,
the following remarks will be centered upon extreme high-
pressure seals, i.e. elements sealing pressures in excess of
50 bar. In view of the different problems entailed by the various media to be sealed, the following remarks are limited
to water applications; fundamentally, however, they are valid
for the whole range of applications.

Figure 1 shows a typical high-pressure seal installed in a
boiler circulating pump that achieved a service life of better
than 20,000 hours under a pressure of $p_1 = 72$ bar and a rubbing
speed of $v_g = 5.4$ m/s without any reconditioning being required.

II. (1) Acquisition of Operational Parameters

So as to permit further analysis of the problems connected with
extreme high-pressure seals, their design criteria will be
treated in detail, special consideration being given to deformation problems.
One major task of all serious seal manufacturers is the acquisition of data on deformation within the seal assembly, i.e.
of inadmissible mechanical and thermal changes of seal gap geometry under given operating conditions; in addition to trials
run on company test rigs, computer-assisted simulations based
on finite elements are used to solve these problems.

Prior to giving any recommendations as to the seal to be used
in highly specialized applications, it is indispensable to analyse carefully all marginal constraints and operating parameters.

(a) Stoppage - Operation:

As mentioned above, mechanical seals are used in connection
with rotating shafts. Up to a certain measure, seal deformation during operation will be neutralized by certain
adaptations of the wearing member, as a rule a hardened
carbon one, due to locally increased surface pressures;
optimum sealing gap configurations will thus be reestab-
lished automatically.

During stoppages, this self-regulating mechanism cannot be
exploited; thus, the seal will be subject to unfavourable
consequences such as detrimental seal gaps, increased leakage, mounting temperatures etc. Service life of badly

designed seals in stop-and-go applications was thus found
to be severely reduced as compared to the same seal in
continuous operation. This fact is particularly important
in connection with stand-by pumps whose seals will be subject to high pressures and, more often than not, to high
temperatures.

(b) Medium:

Contrary to other abrasive and corrosive media, water is
generally thought to be a rather unproblematical and easily
understood liquid. As a rule, this is specially treated,
e.g. demineralized water where pH values may vary within
a wide range and where water purity may change considerably depending upon the operation or plant type used.

Figure no. 2 shows a magnetic filter provided with a sieve
and installed in the circulation cycle of a severely loaded
seal within the secondary system of a nuclear power plant.

So as to achieve satisfactory service lives, particulate
matter will have to be largely excluded. Another aspect
to be considered is the fact that especially high-purity
liquids such as generator cooling water, due to changes
in specific parameters - lessening of surface tension in
connection with unsuitable face - will provoke roughness
and scarring of sliding materials, which would otherwise
be thought attributable to chemical attack (see figure
no. 3).

Choice of suitable sliding materials, their combination
as well as seal geometry have to be adapted to this type
of situation.

(c) Temperature:

Increases in absolute pressure values will generally be
accompanied by concomitant temperature increases. It goes
without saying that, in this case, provision will have
to be made for increased thermal flow towards the seal
which constitutes a heat sink. Essential factors to be
considered in this context are design and dimensioning
of heat buffers between seal and pump, control of thermal
shocks during failures so as to avoid total loss of sealing
properties, the effects of increased thermal expansion on
the entire unit, and the specific behaviour of materials
used in primary and secondary sealing members.

As a rule, care has to be taken to avoid the formation
of thermal layers within and around the seal, and to assure
optimum transfer of any heat generated by means of suitable
heat exchangers.

(d) Installation Space:

Failure of heavy-duty seals is often due to insufficient
coordination among seal and machinery manufacturers, or
else to insufficient knowledge about the requirements to be
met by the various interfaces between the parts of machinery
concerned.

Factors to be considered in this context are the behaviour of supporting flanges and housing sections subject to pressure and temperature variations, cooling medium management around and within the seal, and safe transmission of axial forces.

Figure no. 4 shows the deformation, at a pressure of $p_1 = 158$ bar, of a pump flange serving as a stationary member support. Results were computed according to the finite-element method, and checked by means of appropriate experiments.

(2) Choice of Materials, Typical Forms of Damage

Considering the extreme loads to which high-pressure seals will be subject, choice of materials takes on special importance.

(a) Metals:

Any metal used has to be a rust-proof, high-resistance grade having a low thermal expansion coefficient so as to avoid additional deterioration of seal deformation characteristics under pressure and thermal strain. For this reason, ferritic materials such as chromium steels having at least a 17% chromium content are to be preferred; as a rule, their resistance to chemicals will be sufficient. Figure 5 lists mechanical and thermal characteristics of certain typical materials.

Due to oxygen depletion, prolonged service may produce limited pitting within O ring grooves, a fact which is sometimes enhanced by certain reactions to elastomers present, i.e. to their graphite content.

(b) Sliding Materials:

Seal life and serviceability depend mainly upon the sliding materials chosen. Due to its excellent $p_1 v_g$ values, practically the only combination used is that between hardened metal and carbon.

(i) The carbons used are mainly synthetic, provided with various impregnations chosen according to their mechanical and thermal characteristics. Due to its higher elastic modulus, improved resistance to fracturing, more favourable heat transfer characteristics and better load-carrying characteristics, carbon impregnated with some metal, e.g. antimony, is definitely superior to any resin-based carbon. When choosing a certain type of carbon, low-friction grades are generally to be preferred; overload characteristics, e.g. during emergency dry runs or plant shutdown, have to be taken into consideration. See figure 6 for a summary of physical properties of common carbon materials.

Frequent forms of damage are wear, blistering (figure 7), and loss of impregnation, the last two being at least

E3-38

partially due to thermal overloading which, owing to different thermal expansion gradients attributable to carbon matrix and impregnating material, leads to component separation and face pitting; this will increase leakage losses and reduce wall thickness, thus weakening materials.

Losses of impregnation metal will lead, in proportion to their size, to considerable increases of frictional coefficients which, in turn, may damage sliding materials by thermal stress cracking. Running irregularities due to faulty manufacturing tend to appear above all in large-volume carbon parts and in large-diameter rings. This problem will generally be solved by strict quality control and by introducing design changes permitting smaller carbon parts.

(ii) Hard Alloys:

Due to their excellent wear resistance, corrosion-proof hard alloys are used predominantly. As a rule, these are nickel-based tungsten carbides; for several years now, these materials have been used with excellent results. In comparison to other wear-resistant materials, these alloys may be machined to extremely smooth surface finishes having low R_a and R_t values, a fact which favourably influences carbon wear characteristics (figure no. 8).

The following figure (no. 9) summarizes the technical data of certain commonly used materials.

Typical forms of damage are shown in

Figure no. 10 – Heat strain cracks, i.e. unmistakeable, sharp-edged, straight radial cracks within the sliding face; they are due to the fact that certain stress gradients of the material were exceeded. This type of effect is triggered by thermal overloads resulting from excessive surface pressures and concomitant excessive frictional coefficients, by foreign matter between the sliding faces or by certain types of face deterioration, such as copper deposits.

Figure no. 11 – Erosion /2/, i.e. weakening of the hard alloy structure due to nickel matrix deterioration which facilitates subsequent mechanical abrasion of the remaining tungsten carbide grains
and

Figure no. 12 – Deposits on the hard alloy such as copper smears produced by the medium. This entails considerable increases in frictional coefficients and will finally lead to heat strain cracks and blistering on the carbon surface.

Silicon carbide is an advanced sliding material that, in addition to its extraordinary hardness, wear resistance, and resistance against chemicals, will present, if at all, the above-described types of damage in highly reduced form.

(c) Elastomers

The choice of materials to be used for secondary sealing elements is more restricted.
Due to its indispensable resistance against hot water,
the most widely used elastomer is ethylene-propylene;
if there are high pressures to be sealed, provision
has to be made against its being extruded, however.

Figure no. 13 compares a new ethylene-propylene ring
and a thermally overloaded specimen from a feed pump
after approx. 1 year's service.

Even though hot water seals have already reached considerable service lives of the order of 50,000 hours
of operation, it is to be recommended to change any
elastomer elements every two years during inspections
or subsequent to intermediate servicing work.

A new type of material, perfluoroelastomers, seems to
hold out the promise of considerably extending present
limits of elastomer use. Due to its somewhat astro-
nomical cost, we are still waiting for large-scale
practical use of this material; moreover, one cannot
but think that, especially for hot water applications,
the values promised might be too optimistic.

(3) Seal Design and Applications

(a) Multi-Stage Pressure Reduction:

The traditional way to solve the problem of sealing
high pressures is doing so in multiple stages, i.e.
high pressures are subdivided into several, normally
equal, partial pressures by means of a corresponding
number of individual seals. The resulting load reduction per stage generally increases seal service life;
at the same time, however, it entails increased connecting, monitoring, and maintenance work. Pressures to be
sealed by the various stages are chosen in accordance
with shaft diameters or the seal diameters resulting
therefrom, since deformation depends in a large measure
upon unit size. The following data may serve as rough
guidelines:

Shaft diameter $d_w \leq 150$ mm   Pressure per stage $p_3 =$
                                                    max. 75 bar
Shaft diameter $d_w \leq 260$ mm   Pressure per stage $p_3 =$
                                                    max. 50 bar

In accordance with the philosophy of reducing pressure
by stages, these limits comprise certain safety margins
with respect to possible seal overloads; they are not
to be interpreted as maximum values to be achieved with
a single-stage seal.

As a rule, pressures are distributed by means of throttle
tubes (capillaries) within the buffer water supply cir-
cuit of the seals or else by means of separate buffer
pressure devices. If one seal fails, the pressure ex-
erted upon the remaining seals will increase accordingly;

every seal will be so designed as to be able to
shoulder, for a certain time, total pressure during
start-up or in an emergency.

Figure 14 shows a triple tandem seal designed according
to the functional philosophy described above, and installed in the primary-cycle circulating pump of a
boiling water reactor.

(b) Single-Stage Seals:

Increased capital outlay and high monitoring requirements entailed, as already mentioned, by multi-stage
seals as well as certain problems encountered in properly splitting up pressures, is inducing more and more
customers to demand single-stage seals.

However, this type of design requires increased attention to deformation and cooling problems.

Some essential parameters:

- proper transfer of any frictionally generated heat
  by means of optimum cooling system design and possibly by distributing devices,

- almost total hydraulic balancing of the seal while
  maintaining sufficient interface widths so as not
  to exceed sliding material resistance parameters,

- assuring simplicity when designing seal elements so
  as to keep deformation within manageable limits,

- exclusion, to a wide extent, of abrasive particles
  within the medium.

Present state-of-the-art seals managing single-stage
pressures of p_1 = 160 bar and $p_1 \cdot v_g$ values of 3600 bar
m/s will certainly not prove to be last word in seal
design.
Picture 15 and 16 are showing a basic diagram of such
single stage high pressure seals.

(c) Thermo-Hydrodynamic Design:

Proper functioning of any heavy-duty seal depends upon
control of frictional parameters within the seal gap.
Knowledge of these parameters and proper choice of
sliding materials almost invariably lead to these materials being equipped with thermo-hydrodynamic grooves
(see figure no. 17) /3/.

This feature has become indispensable, since friction
coefficients will be decreased as pressures increase,
and since thermo-hydrodynamic elements are inherently
self-regulating, i.e. sealing action will increase together with friction just as groove area cooling will
improve due to increased liquid flow into the seal gap;
moreover, such seals will be more resistant to short-
term in-operation overloads.

It has to be said, however, that medium purity is more important when thermo-hydrodynamical seals are used; it is easier for particulate foreign matter to be flushed into the seal gap. Seal life may thus be considerably shortened; even immediate failures might occur.

(d) Supporting Rings:

Based on knowledge about deformation of the various materials combined in a given seal, particular attention has to be paid to the properties of carbon, the stability of which is vastly inferior to that of hard alloys and steel. So as to decrease deformation, carbon sliding rings are provided with radial supporting rings when used with pressures > 50 bar. Choice of a suitably tempered material is essential, since badly chosen rings will come loose when loaded and entail further damage. For this application, corrosion-resistant ferritic steel grades are to be recommended, too.

So as to prevent the dynamically loaded elastomer rings required in every seal from being extruded axially, backing rings will be used in addition to rings having high Shore hardness values. Suitable materials - having proper sliding characteristics, low water adsorption, high resistance against thermal and chemical attack, and low flow rates, to name just a few properties - are just as important as proper processing of such materials.

When PTFE-based materials are used, whether with filler or fillerless, their unfavourable expansion coefficients within the $0^{\circ}C$ to $70^{\circ}C$ range will bring about recurring problems if proper processing temperature is neglected.

III. Future Outlook

Due to the continuing evolution of sliding materials - carbon grades as well as hard alloys - and owing to the fact that increasing attention is being given to deformation problems within and near seals, in-operation single-stage pressures of up to some 160 bar may now be realized. This can only be achieved, however, if there is close cooperation with pump producers, since seal function depends critically upon marginal pump constraints. Further work will be necessary to permit better understanding of discontinuous temperature distributions such as thermal shocks so as to provide for efficient counter-measures.

<u>References</u>

/1/   E. Mayer: Mechanical Seals (Axiale Gleitringdichtungen)
                  6th edition
                  VDI-Verlag Düsseldorf 1977 (in German)

/2/   E. Mayer: Experiences Gained in the Operation of Heavily Loaded
                  Axial Seals Used in Power Plants (Betriebserfahrungen
                  mit hochbelasteten Gleitringdichtungen in Kraftwerken)
                  6. Internationale Dichtungstagung
                  Dresden, April 1978 (in German)

/3/   E. Mayer: High Duty Mechanical Seals for Nuclear Power Stations
                  5th Int. Conf. on Fluid Sealing
                  Coventry/England 1971, Paper A5

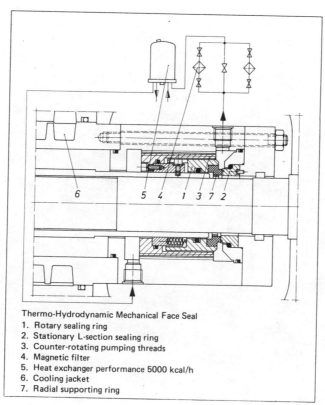

Thermo-Hydrodynamic Mechanical Face Seal
1. Rotary sealing ring
2. Stationary L-section sealing ring
3. Counter-rotating pumping threads
4. Magnetic filter
5. Heat exchanger performance 5000 kcal/h
6. Cooling jacket
7. Radial supporting ring

Fig. 1.   Burgmann high pressure seal of a boiler circulating pump

Fig. 2. Magnetic filter used within the circulation cycle of a heavy-duty
seal to be employed in a nuclear power plant (medium - feed water)

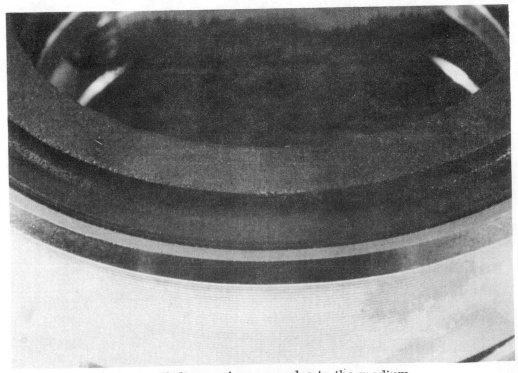

Fig. 3. Sliding carbon wear due to the medium

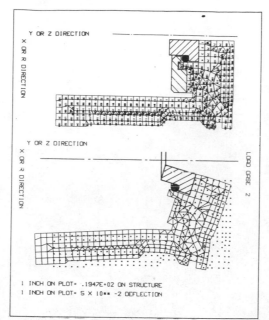

Fig. 4. Deformation of a floating member support for a medium pressure of $p_1 = 158$ bar, finite-element computation

Physical Properties of Sliding Face Materials (Average Values)

Steel Grades

Material	Material No.	DIN Designation	Brinell Hardness (N/mm²)	0.2% Expansion Limit (N/mm²)	Elastic Modulus E (10⁴ N/mm²)	Expansion Coefficient (10⁻⁶/K)	Thermal Conductivity Coefficient λ ($\frac{W}{mK}$)
Chromium steel	1.4122	X35 CrMo17	2250-2750	600	21.3	10.5	29.3
Chromium/nickel steel	1.4057	X22CrNi17	2250-2750	600	21.0	10.0	25.2
Chromium/nickel steel	1.4313	G-X5 CrNi13 4	2300-3000	650		12	25.2
CrNiMo steel	1.4460	X8 CrNiMo27 5	1900-2300	500	21.0	11.5	14.6
CrNiMo steel	1.4571	X10 CrNiMoTi18 10	1300-1900	230	20.3	16.5	14.6

Fig. 5. Material parameters, steels

Physical Properties of Sliding Face Materials (Average Values)

Synthetic Carbons

Material	Compressive Strength δ_D (N/mm²)	Density ρ (g/cm³)	Elastic Modulus E Deflection (10⁴ N/mm²)	Expansion Coefficient α (10⁻⁶/K)	Thermal Conductivity Coefficient λ ($\frac{W}{mK}$)
Carbon resin impregnated	230	1.8	2.7	4.3	10
Carbon antimony impregnated	310	2.5	2.1	4.7	13
Electro-graphite resin impregnated	150	1.8	1.1	4.3	65
Electro-graphite antimony impregnated	170	2.5	1.6	4.5	65

Fig. 6. Material parameters, sliding carbons

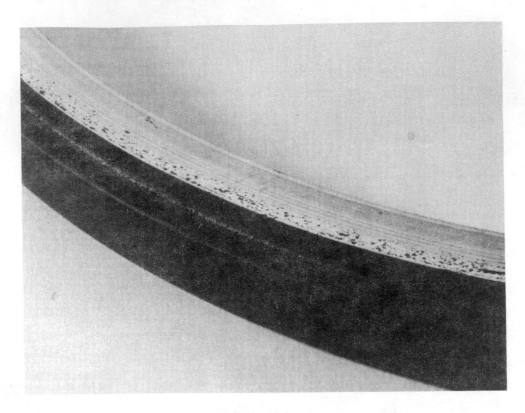

Fig. 7. Carbon sliding ring showing blistering

Arithmetically Averaged Roughness Values
of Sliding Face Materials

Hard alloy tungsten carbide	0.015 μm
Hard alloy silicon carbide	0.05 μm
Special cast iron	0.2 μm
Carbon	0.3 μm
Ceramic material	0.35 μm

Fig. 8. $R_A - R_t$ values of standard sliding face materials

Physical Properties of Sliding Face Materials (Average Values)

Metal Carbides

Material	Hardness (N/mm^2)	Density ρ (g/cm^3)	Elastic Modulus E $(10^4\ N/mm^2)$	Compressive Resistance σ_D (N/mm^2)	Expansion Coefficient α $(10^{-6}/K)$	Thermal Conductivity Coefficient λ $(\frac{W}{mK})$
Tungsten carbide	15000	14.8	60	5000	4.5	82
Silicon carbide	25000	3.1	41	3500	4.3	84
Compare:						
Aluminium oxide (99.7%)	22000	3.9	35	4000	7	26

Fig. 9. Material parameters, hard alloys

Fig. 10. Tungsten carbide ring showing thermal strain cracks

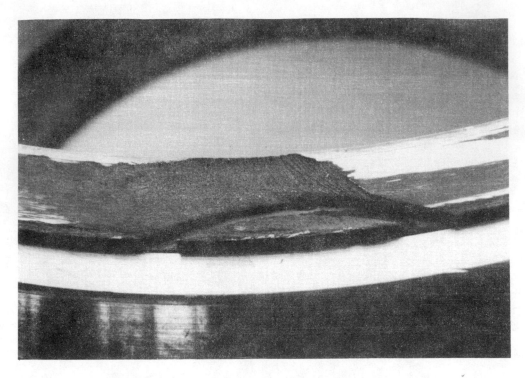

Fig. 11. Erosion within the sliding face region of a tungsten carbide ring

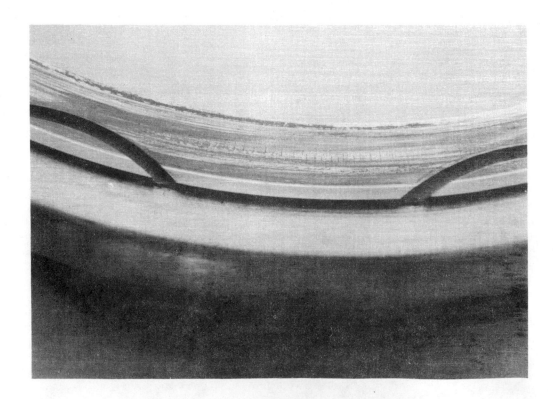

Fig. 12. Hard alloy ring used in a feed pump seal and showing copper deposits and thermal strain cracks

Fig. 13. No. 1 is a new o-ring, no. 2 and no. 3 show
those that have been thermally overloaded.

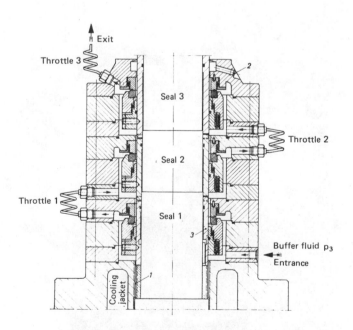

Fig. 14. Triple tandem seal of main cooling medium pump

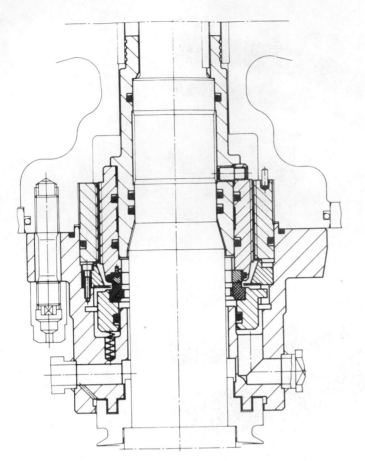

Fig. 15. Basic diagram of a high pressure
         seal to be used in a nuclear pilot
         pump

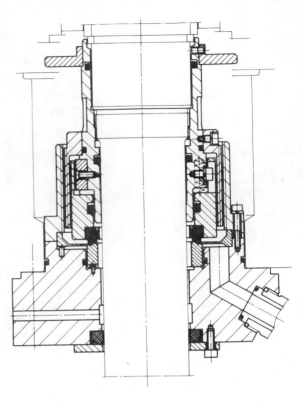

Fig. 16. Sealing design diagram of a
         reactor water clean up pump

Fig. 17. Seal parts provided with thermo-hydrodynamic circulation grooves

HYDRAULIC SEALS IN THE MINING INDUSTRY

H. Hopp

Martin Merkel KG, Federal Republic of Germany

and

M. J. Harwood

FTL Company, U.K.

Held at the University of Durham, England.

Conference sponsored and organised by BHRA Fluid Engineering

©BHRA Fluid Engineering, Cranfield, Bedford MK43 0AJ, England.

HYDRAULIC SEALS IN THE MINING INDUSTRY

'Safety first' has always been the watchword of the mining
industry.

Among the many problems that the seal manufacturer has to contend
with are:-

> non-flammable fluids
> high hydraulic pressures
> negligible stroke and
> cylinder wall deformation

Non-flammable fluids, in general, fall into two categories - the
so-called HSA fluid, a water/oil emulsion, and the HSD fluid which
is a fully synthetic hydraulic fluid most commonly of the phosphate
ester type.

No general introduction shall be provided, but we shall mention 3
specific problems we have encountered in underground applications.

1. Seals for roof supports -

    Roof supports, or to use their original name, pit props, are
    either single, or nowadays, a series of double acting hydraulic
    cylinders used to keep the roof of the coal seam from meeting
    the floor in the longwall-mining application.

    Because of the need to keep the equipment as physically compact
    as possible, and also the very high load that the support is
    subject to, hydraulic pressures of between 500 and 750 bars are
    not uncommon.

    In an effort to reduce the equipment's weight, and to facilitate
    easy transportation, the mechanical properties of the material
    used in the cylinders are made full use of. Contrary to normal
    hydraulic practice, larger than normal deformations are acceptable.
    This is particularly so in the case of the cylinder barrel.
    Because of variable loading and also actual tectonic movements,
    the supports must be able to withstand a high rate of bending.

The wear resulted from the fact that the rubber portion of the seal had not been able to retain a film of lubricant and had been running dry, whereas the fabric portion had.

Due to short, fast, strokes, that the piston made, the heat caused by the friction of the dry rubber caused further expansion in the rubber and premature wear and failure.

The seal was re-designed, as shown, to incorporate a rubber/fabric sealing portion along its entire length. Since then, no further trouble has been experienced.

Fig. 1 Telescopic Prop:the photograph shows simulation on a testing apparatus of underground bending

Figure 1 shows a telescopic roof support undergoing a test simulating the type of bending that it may experience underground. In spite of this high bending moment, the seals in the assembly must not fail after a total extension of several thousand metres, equivalent to a two year service life.

In the present generation of roof supports, the so-called COMPACT-seal is used. Figure 2 shows the general arrangement of the two stage telescopic roof support under test. You will note that the seals are all equipped with a back-up, or anti-extrusion ring, (manufactured from a hard plastic material), preventing the softer sealing portion of the seal extruding.

Extrusion occurs due to the seal being forced by the hydraulic pressure into the clearance gap, the clearance gap being not only due to the normal manufacturing tolerances and clearances, but also, and this being more critical, to the elastic deformation of the cylinders caused by high pressures and external loads.

Figure 3 shows an enlargement of the double acting piston seal. The dynamic sealing element of the seal is an elastomer impregnated fabric.

Pure elastomer seals tend to stick to cylinder walls after long periods of high hydraulic pressures and virtually no stroke. This results in jerky operation, tearing and subsequent rapid failure. The advantage of a strong fabric based piston seal, which does not exhibit these properties, become self-evident, when a slow push movement of about 20 mm/min. is necessary.

Further explanations:-

The outer surface of the seal, under hydraulic pressure, is pressed against the cylinder wall. With very low traverse speeds, there will be no uniform expansion at the outer surface of the elastic rubber ring because of the static friction and the slightly variable elasticity of the material. Because of this, tensile stresses are developed in the surface of the seal. When this tensile stress exceeds the value of static friction, then this results in regaining of the original form from the state of deformation (dynamic friction). This process always takes place throughout the complete stroke at a very high frequency and leads to a relatively quicker ageing and shortening of the life of the seal. The fabric reinforcement of the outer surface of the seal prevents this expansion occurring. Furthermore, the pattern of the fabric layer on the sliding surface helps to build up a lubricant film.

The COMPACT type of seal offers a high rate of sealing, even under variable deformations, due to the high preload of the seal lip caused by the radial loading of the seal.

The high friction at zero pressure, compared with that of lip seals, is not important in this application.

For larger diameters (approx. 5" and above), the piston seal ring shown for example in Figure 2, can be snapped over the piston for assembly. Naturally, this depends upon the radial width of the seal. Rings with smaller diameters, and smaller widths, can also be mounted in this way. But here, one has to take care. The radial width chosen must be as large as possible because one has to take into consideration, not only the tolerances of the metal parts which are normally very "tight", but also deformation under pressure as well as extreme lateral loads as shown in Figure 1.

The fabric dynamic sealing surface always allows sufficient lubricant between the seal and the cylinder surface. This is very important because water, sometimes in poor condition, is used as hydraulic fluid.

A COMPACT seal can also be used for sealing the piston rod.

A seal with several sealing edges is shown in Figure 2. There again, a fabric reinforced sealing allows for retention of a lubricant film, necessary for smooth slow movement and a long service life.

2. Seals for hydraulic loaders -

Figure 4 shows a hydraulic loader in action underground. Operating conditions for this type of underground machine do not differ from those of normal earth moving equipment. The construction of the cylinders is thus similar to conventional hydraulic cylinders. However it is fairly common to use hydraulic fluids of the phosphate ester type, (although this is more common in Europe than in the U.K.).

Figure 5 shows the design of the main cylinders from this machine. A COMPACT piston seal is used. To ensure compatability with the hydraulic fluid and also to gain maximum benefit from its properties, its elastomer parts are made of fluorocarbon elastomer (VITON) and the fabric part from VITON proofed asbestos cloth.

The asbestos fabric, having a high material stability, is well suited for the long tempering process at high temperature necessary for the correct manufacture of these type of seals. Cotton fabric, the normal fabric component of this type of seal, will not withstand this high temperature process.

Tests by the Laboratory of Steinkohlen-Bergbau, Essen, on a loader, running with a phosphate ester fluid, showed that no decrease in sealing efficiency was experienced after 70,000 double strokes of this cylinder. Moreover, wear was undetectable.

Many of these loaders are successfully operating now for many years, proving that the design of the cylinder and its parts are well suited for this arduous application.

3. Tension cylinder on a mono-rail system -

The traction rope of this mono rail system, operating underground has to be kept in permanent tension. Figure 6 shows the power unit with the tension cylinder. Figure 7 shows the type of piston arrangement in the hydraulic cylinder. Originally, a nitrile elastomer seal with lateral layers of fabric was used. The seal operated at 300 bar max and because of its function, only had a very small length of stroke. After three months' operation, the seal exhibited considerable wear on the rubber portion but very little on the fabric part. The diagram compares the seal in the new state and after 3 months operation.

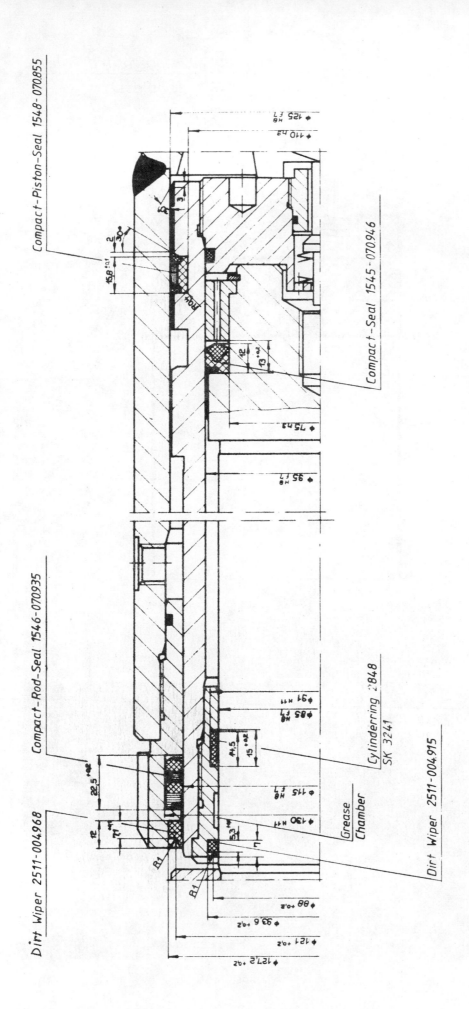

Fig. 2 Seal for hydraulic prop.

J3-27

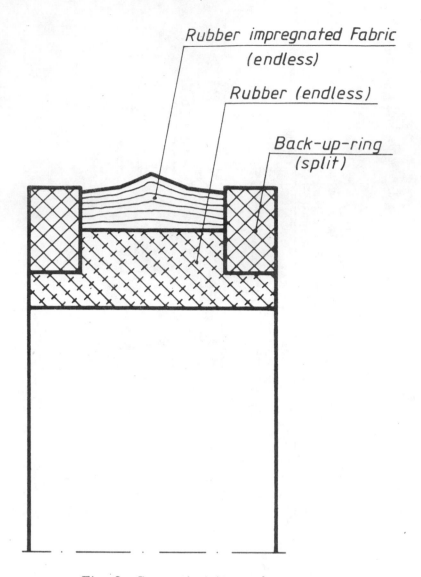

Rubber impregnated Fabric
(endless)

Rubber (endless)

Back-up-ring
(split)

Fig. 3   Compact piston seal.

Fig. 4   Hydraulic charger.

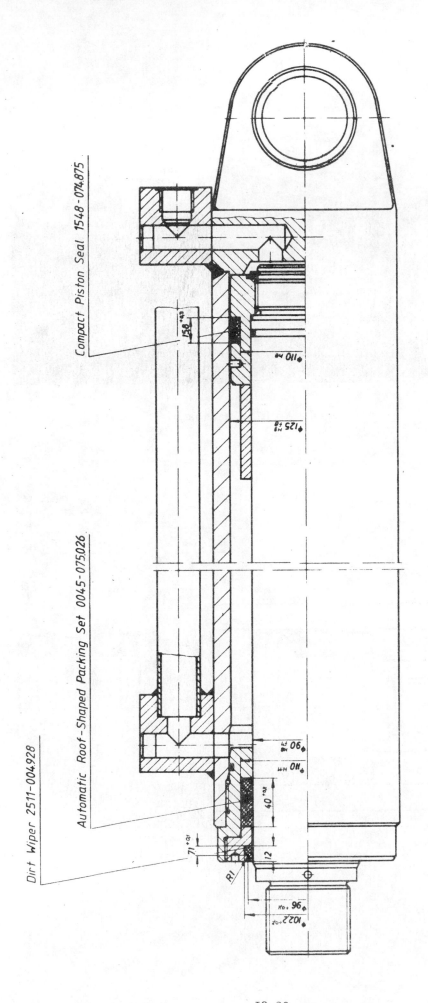

Fig. 5 Seal for hydraulic cylinder.

Compact Piston Seal 1548 - 074.875

Dirt Wiper 2511-004.928

Automatic Roof-Shaped Packing Set 0045-075.026

J3-29

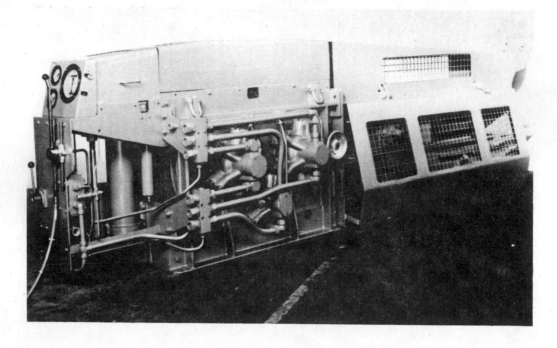

Fig. 6 A lifting machine.

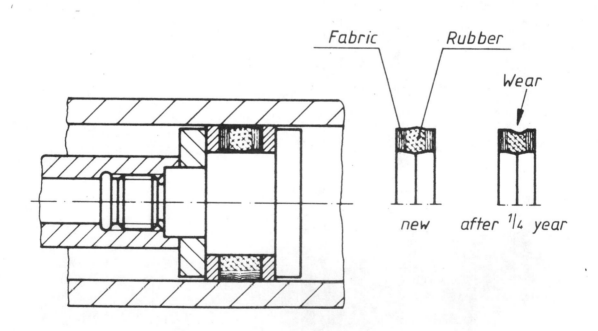

Fig. 7 Tension cylinder.

OLD   NEW

Fig. 8   Worn out was the rubber portion at the bracing cylinder, after a quarter of a year's operation of the seal.   The new ring at the right side.

Discussion & Contributions

SESSION A : MECHANICAL FACE SEALS - I

Chairman: Mr. R.A. Grimston,
Flexibox Ltd., U.K.

Papers:

A1   Diametral tilt and leakage of end face seals with convergent sealing gap.
E. Metcalfe, N.E. Pothier and B.H. Rod, Atomic Energy of Canada Ltd., Canada.

A2   Dynamic characteristics of face seals.
M. Kaneta, and M. Fukahori, Kyushu Institute of Technology, and F. Hirano, Kyushu University, Japan.

A3   An assessment of factors affecting the response of mechanical seals to shaft vibration.
R.T. Rowles, and B.S. Nau, BHRA Fluid Engineering, U.K.

A4   Preliminary experiments with slot fed hydrostatic face seals.
I.T. Laurenson and J.P. O'Donoghue, Northern Ireland Polytechnic, U.K.

Notes:

1.   The author in bold print presented the paper.

2.   Errata on Papers A2, A3 and A4 is provided overleaf.

ERRATA

PAPER A2

Page A2–23

Table 3 – the following should be noted:

upper figures : experimental runs
lower figures : percentage of occurrences (colum summation is 100).

PAPER A3

Page A3–31

The equation should read:-

$$\pi\, r\, \rho\ \ A\ \ \left(1 + \frac{1}{i^2}\right)\ \ddot{x}\ +\ \frac{\pi\, E\ I}{r^3}\ \left(1 - i^2\right)^2\ x\ =\ 0$$

Page A3–32

The first three equations should read:-

$$\text{flexural stiffness} = \frac{\pi E I}{r^3}\ \left(1 - i^2\right)^2$$

$$\omega^2 = \frac{E}{\rho\, r^2}$$

$$\text{stiffness} = \frac{2\,\pi\, E\ l\ w}{r}$$

Section 2, line 5 should read:-

".... is 4×10^{12} N.m^{-1}."

Section 4, last line should read:-

"2×10^9 N.m^{-1}."

Page A3–36

Fig. 1:- The top figure of 10% in the right hand column should read 90%.

PAPER A4

Please note the pages listed should be as follows:-

A4–44   should be   A4–46
A4–45   should be   A4–44
A4–46   should be   A4–45
A4–47   should be   A4–48
A4–48   should be   A4–47.

Authors' replies follow the individual questions.

PAPER A1

Diametral tilt and leakage of end face seals with convergent sealing gap.

E. Metcalfe, N.E. Pothier and B.H. Rod

Atomic Energy of Canada Ltd., Canada.

R.S.L. WEIR, ESSO PETROLEUM CO. LTD., U.K.

Could you please say what were the face materials used? Was the face separation measured with the seals running and if so how was it done?

Authors' reply

Face materials for the "larger" seals were tungsten carbide against a nickel-chrome alloy, and for the "smaller" seals tungsten carbide against 410 stainless steel. In each case the tilt load just to cause rubbing contact, with the seals running, was easily sensed by hand because the rig body was freely mounted in bearings.

D. HUHN, GUSTAV HUHN AB, SWEDEN

Referring to Fig. 7 of the paper (showing the test rig)- the test seal seems not to have any springs.

The lack of springs might contribute to the tilting of the seal face, especially at higher degrees of balancing the hydraulic pressure. I certainly cannot agree that the springs have importance for supplying an initial force at low pressures.

Authors' reply

No springs were used and none were needed in the tests. With the seal back pressure system, seal faces could be forced together by any desired load to give any desired balance ratio. Tilt load was applied independently by another pressure system. For the purposes of measuring the tilt load just to cause rubbing contact, the presence or absence of springs has no significance.

R.T. ROWLES, BHRA FLUID ENGINEERING, U.K.

Your paper gives an interesting discussion of the effect of tilt on seal performance. However, I am concerned at the magnitude of the initial convergences quoted. Although there is of course no direct comparison, a seal with a circumferential wave profile of 3 μm will leak copiously and I would expect a seal with the convergences quoted here to leak copiously. It is not surprising that any tilt large enough to counteract such convergence results in poor performance or failure.

Have the authors been able to determine whether the convergence is maintained during running? Have they considered the behaviour of a seal with smaller radial convergence, of the order of 1 μm?

Have the authors measured the residual circumferential waviness of their seals? If so, what was the profile? If the profile has typical values, around 0.5 - 1 μm peak to peak wave amplitude. Hydrodynamic effects would be expected.

Authors' reply

For full film lubrication of a conical-convergent seal, only sufficient convergence to overcome manufacturing and deflection inaccuracies is required. Experience shows that for the test seals and pressures, 1 μm convergence is normally sufficient. However, the static tilt load just to cause rubbing of such a seal would vary greatly with angular position because of the deforming effect of pressure on the seal rings. Ideally the ring shapes and 0-ring positions were designed for zero seal face deformation with pressure. In practice nothing is quite ideal. A minimum value of 2.65 μm convergence was chosen to avoid this problem. Larger values were then tested to establish the predicted independence of tilt load, not because leak rates up to 4 ℓ/min are desirable.

No significant difference between static and dynamic leak rate was measured for any of the untilted test seals, showing yes, that convergence was apparently not affected by running, and no, circumferential waviness was not sufficient to cause significant hydrodynamic effects. In all cases there was no initial or final (residual) circumferential waviness of peak to peak amplitude more than 5% of the lapped-in convergence.

Dynamic characteristics of face seals.

M. Kaneta and M. Fukahori

Kyushu Institute of Technology, Japan

and F. Hirano

Kyushu University, Japan

A.C. PIJCKE, NETHERLANDS MARITIME INSTITUTE.

You did not state that the sealing performance is influenced by the two phase flow due to the dynamic behaviour of the sealing surfaces. Could you tell what kind or different kinds of two phase flow you have observed since each kind has a different heat transfer coefficient.

Authors' reply

We observed mainly three types of flow pattern in the fluid film between sealing surfaces in this experiment as follows: full liquid film, full air film and incontinuous liquid film containing a gas band. In this experiment, however, we have not been able to classify the detailed patterns of the two phase flow. However, one of us and co-workers have observed microscopic changes in the behaviour of the gas - liquid interface and flow patterns in the liquid film between an optical flat vibrating in the normal direction or a rotating plate and a stationary plate. (Refs. 1-3). We have found different kinds of two phase flow. The relationship between these flow patterns and the sealing performance will be presented in the future.

References

1. Saki, K., Hirano, F., Sakai, T. and Shimogama, T.: "The behaviour of oil film between the static and the vibrating plate (1st report)". Journal of Japan Society of Lubrication Engineers, **20**, pp. 552-559, (1975).
2. Saki, K., Hirano, F. and Sakai, T.: "The behaviour of oil film between the static and the vibrating plate (2nd report)". Journal of Japan Society of Lubrication Engineers, **20**, 8, pp. 560-566. (1975).
3. Hirano, F., Kubo, S., Saki, K. and Ikeda, Y.: "The investigation of gas-liquid 2-phase flow between rotating and stationary plates". Journal of Japan Society of Lubrication Engineers, **20**, 8, pp. 567-573. (1975).

M.T. THEW, SOUTHAMPTON UNIVERSITY, U.K.

The Z parameter, which is significant in governing air ingestion, contains the surface tension in the denominator. When using water the surface tension may easily be varied over a range of say 3:1 by adding surfactants. This would be a valuable addition to the experiments. Has this been done or will this experiment be carried out?

Authors' reply

The experiment you suggest has not been done directly. In the case of distilled water, the temperature rises reached about 90°C, so that it was difficult to distinguish the types of fluid films i.e. the inhalation of air or the boiling of water. Fig. 1 overleaf shows the Z-G plots for water. The dotted line corresponds with the limit line between the leaking and the inhalation obtained for oils.

In the case of oils, however, one of the authors has already pointed out the important role of the surface tension, especially its variation caused by the temperature gradient on a heating surface, i.e. the Marangoni effect, on the sealing performance of a mechanical seal. (Ref. 4 of the paper).

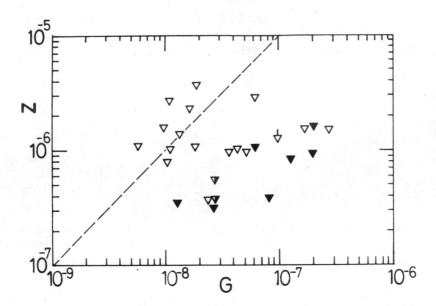

Fig. 1   Relation between Z and G (Coaxial seal) for water.

P.G. MOLARI, UNIVERSITY OF BOLOGNA, ITALY.

What is the quantitative influence of the silting up (or molecular clogging) effect for different oils in the presence of different levels of vibration?

At low pressures and with narrow gaps my experience is that this phenomenon is relevant and unpredictable and it requires a deep study on the gathering mechanism of polar molecules (stearic acid and so on). Perhaps it is partially the reason for the spread of the results?

Authors' reply

We agree with your comments concerning the significance of the interfacial phenomena. We are now planning some experimental investigations on the matching effect of numbers of carbon atoms between fatty acids and hydrocarbons as solvents (Ref. 4 and 5 below), on the sealing performance of mechanical seals.

References

4. Askwith, T.C., Crouch, R.F. and Cameron, A.:   "Chain length of additives in relation to lubricant in thin film and boundary lubrication", Proc. Roy. Soc. London., **A291**, pp. 500–519, (1966).

5. Hirano, F.: "Effects of chain length of fatty acids added to hydrocarbons on surface viscosity, scuffing load, and cavitation erosion". Frontiers in Lubricant Technology, Naples, Session D, (April, 1978).

B.S. NAU, BHRA FLUID ENGINEERING, U.K.

Are both the high frequency vibration and the Marangoni effect necessary together to explain the film character and air ingestion?

Is it not possible that the high frequency vibration is a red herring since it is associated with increased air content of the film and the damping will then be reduced?

Authors' reply

At this present stage, we consider that both the high frequency vibration and the Marangoni effect are necessary to explain the sealing performance. However, we can only say that we could not have explained the origin of the vibration except in the case of Type D, which may correspond with the natural frequency of the testing machine. The relationship between the vibration and the behaviour of fluid will be made clear by our next set of experiments.

D. HUHN, GUSTAV HUHN AB, SWEDEN.

The paper does not refer to the pressure of the oil under which the tests were carried out.

As different amounts of gas can be dissolved in the fluid at different pressures and are again gasifying at lower pressure could you please specify the amount of pressure for each of the tests reported?

Authors' reply

All experiments were carried out under atmospheric conditions on both the inside and the outside of the sealing surface. Oil in the glass pipe gave small gravitational pressure but its effect was negligible, so that there was practically no pressure difference. There was also no notable difference in the sealing performance between the usual oil and the de-gassed oil as the liquid to be sealed. Therefore, we consider that the influence of the pressure difference is not appreciable upon the phenomena we have observed.

R.S.L. WEIR, ESSO PETROLEUM CO. LTD., U.K.

Are you familiar with the work of Orcutt and M.T.I. who used a similar test rig with an optical flat glass face? In that instance it was claimed that a substantial temperature rise was observed at the seal faces. Did you notice any major face temperature rise?

Authors' reply

The temperature rises during oil leakage were a little higher than during inward pumping, i.e. inhalation of air, due to its cooling effect. However, as has been pointed out in this paper, any major temperature rise and its notable effects were not observed except in the case of water.

PAPER A3

An assessment of factors affecting the response of mechanical seals to shaft vibration.

R.T. Rowles and B.S. Nau

BHRA Fluid Engineering, U.K.

D. HUHN, GUSTAV HUHN AB, SWEDEN.

Besides the frequency of vibrations the amplitude is very important to get hold of. In many applications, i.e. in propeller shafts in ships, vibrations are not only mechanically generated but also hydraulically by the sea and by the ship propeller especially on an aft stern tube seal. Seal manufacturers often lack this kind of information on existing frequencies and amplitudes to provide a proper seal solution.

These conditions can become severe up to shock loads to be be supported by the seal. What type of vibrations are you referring to and have you taken into account these extreme values in your calculations?

When heavy vibrations occur, 0-rings are not normally used as a secondary seal, but bellows of synthetic rubber or metal and rubber sleeves are preferred. Have you taken into account these forms of secondary seal?

Authors' reply

The vibrations referred to in the paper are those that can be modelled by a sine wave of known amplitude and frequency. The solution technique used can be extended to model shock loads and these may be considered at a later date.

The authors are aware that 0-rings may not be used when heavy vibrations occur, however, 0-rings are the most common type of secondary seal.

H.K. MULLER, UNIVERSITY OF STUTTGART, FEDERAL REPUBLIC OF GERMANY.

When considering the different influences on the vibration of the floating ring of a face seal in your conclusion, you have stated that the film forces are in excess of the axial 0-ring damping forces. Now the fluid film forces easily can be calculated for the pushing period of the vibrating motion of the seal, that is when the faces move to close the clearance. On the other hand when the faces are moving apart the film pressure may either cavitate or, as other investigators have shown, transmit appreciable tensile stresses at least for a limited time.

How in your vibration analysis did you take into account cavitation and the ability of liquids to transmit tension forces between the seal faces?

Authors' reply

This is a problem which requires serious consideration even though much of the work on tensile stresses in fluid films has been shown not to apply to a mechanical seal configuration (Ref. 1 below). We have assumed that hydrodynamic pressure generation occurs with cavitation which occurs instantaneously. Time dependent effects are a refinement which it may be necessary to include at a later date.

Reference
1. Savage, M.D.: "Cavitation in lubrication". J. Fluid Mech., **80**, 4, pp. 743-767. (23 May, 1977).

J.P. O'DONOGHUE, NORTHERN IRELAND POLYTECHNIC, U.K.

I draw your attention to the statement in the first paragraph on p. A3-33. Surely the stiffness is not dependent on the term dh/dt, this being a damping term. It is important if a seal is to be effective that the ring stays in contact with the housing, consequently in designing a mechanical face seal it is essential to ensure that the axial film stiffness and damping are 2-3 orders of magnitude higher than those of the secondary seal. In choosing a practical design you have confirmed this to be the case, but the point is not made in the paper, that this is essential for designing a seal.

Authors' reply

You correctly draw attention to the erraneous comment about dh/dt and stiffness. The paragraph appears to have become garbled during editing. It was intended to point out that the stiffness value 8×10^8 N.m^{-1} is valid only for small perturbations of film thickness about the value used since the stiffness is a function of film thickness. The damping due to dh/dt is dealt with in the second paragraph.

R. METCALFE, ATOMIC ENERGY OF CANADA.

You have provided a useful perspective of the influences on a typical end face seal. However, your treatment of the fluid film would be enhanced by reference to the recent work of Etsion (Refs. 1, 2, and 3 below). Of particular concern here is the significant fluid film rocking (tilting) mode. The expression obtained by you for "stiffness" $\partial W/\partial \psi$, is not tilting stiffness in the conventional sense but is just one component (hydrodynamic) of the cross effect of tilt on axial force. In full film liquid lubricated seals there is no direct hydrodynamic tilting stiffness because the induced moment is at 90° to the axis of tilt. However, there will generally be a hydrostatic tilting stiffness, estimated (Ref. 4) to be 35,000 N.m/rad for your sample seal at nominal 74% balance ratio and 1 MPa pressure drop. Test results for hydrostatic tilt resistance have been reported recently (Ref. 5) by this discusser, along with some measurement of hydrodynamic effects.

Could you therefore, please clarify your comments on fluid film effects in the tilting mode, particularly as related to your prime objective of assessing the relative importance in various modes of seal vibration?

References

1. Etsion, I.: "Hydrodynamic effects in a misaligned radial face seal". ASME Paper 78-LUB -12. (October, 1978).

2. Etsion, I.: "The accuracy of the narrow seal approximation in analysing radial face seals". ASLE Paper 78-LC-2B-2. (October, 1978).

3. Etsion, I.: "Radial forces in a misaligned radial face seal". ASME Paper 78-LUB-13. (October, 1978).

4. Metcalfe, R.: "Predicted effects of sealing gap convergence on performance of plain end face seals". ASLE Trans., **21**, 2, (April, 1978).

5. Metcalfe, R., Pothier, N.E. and Rod, B.H.: "Diametral tilt of end face seals with convergent sealing gaps". Proc. 8th Int. Conf. on Fluid Sealing. Cranfield, U.K., BHRA Fluid Engineering. Paper A1, pp. A1-1 - A1-14. (September, 1978).

Authors' reply

We would like to thank you for bringing the very recent work of Etsion to our attention.

With regard to the comments on the rocking and tilting stiffness we agree that the value calculated in the paper is an axial component due to hydrodynamically generated resistance to tilting.

In the angular mode the ring dynamics will, in general, depend on a moment:-

$$M = g \left(h_o, \frac{\partial h}{\partial t}, \psi, \frac{\partial \psi}{\partial t} \right)$$

Using a similar analysis to that carried out in sections 7 and 9 of the paper the following angular stiffness values have been calculated for a small change in axial position and a small change in angular misalignment respectively:-

$$\frac{\partial M}{\partial h_o} = 2 \times 10^8 \text{ N.m/m} \qquad \frac{\partial M}{\partial \psi} = 1.5 \times 10^7 \text{ N.m/rad.}$$

These values are similar to those calculated in the paper and do not affect the authors comments. They do however show that the hydrodynamic component of tilting resistance is more important than the hydrostatic component, calculated by Dr. Metcalfe to be 3.5×10^4 N.m/rad for the seal considered.

Preliminary experiments with slot fed hydrostatic face seals.

I.T. Laurenson and J.P. O'Donoghue.

Northern Ireland Polytechnic, U.K.

R. METCALFE, ATOMIC ENERGY OF CANADA.

Hydrostatic seals have been used with some success in nuclear primary circulating pumps. However, as you point out and as Canadian experience shows, silting or other interference by dirt is difficult to eliminate. Your seal is a worthwhile addition to the collection of "dirt resistant" hydrostatic seals, which now includes in-face labyrinth channels (Ref. 1 below), conical and stepped-face seals (Ref. 2). Is the slot controlled seal more "resistant" than these? Does erosion of the face opposite the discharge from the pocket cause problems, as with other pocket-type seals?

On testing of the single-pocket seals, were no tilting (rocking) problems encountered? Other results (Ref. 1) show single-pocket seals to be impractical, while six pockets were required for close-to-optimum tilt stability.

References
1. Pothier, N. and Rod, B.H.: "Development testing of a pocket-type hydrostatic rotary end face seal in high pressure water". ASLE Paper 78-LC-3B-2. (1978).
2. Bell, R.P.: "Comparison of off-design performance of various hydrostatic seals". Proc. 5th Int. Conf. on Fluid Sealing. Cranfield, U.K., BHRA Fluid Engineering. Paper E5, pp. E5-65 - E5-72. (1971).

Authors' reply

There is at present, insufficient evidence to determine whether or not the slot controlled seal is more resistant to silting than the other dirt resistant seals you mention, but it does appear to be self cleaning when there is rotation, if the silt is not too strongly adherent to the slot surface. No evidence of silting has been observed during prolonged running in conditions which produce rapid silting when there is no rotation. Additionally, the multi-pocket slot controlled seal has the advantage of greater stiffness under tilting conditions than the conical and stepped face seals.

No erosion of the face opposite the discharge from the pocket has been observed, but the test durations to date have been short compared with the expected lives for hydrostatic face seals.

The only tests carried out on single recess seals, to date, have been short duration tests to investigate hydrodynamic effects. No tilting problems were encountered during these tests, but it is agreed that single recess seals are inferior to multi-pocket seals under tilting conditions. Design analysis confirms that six pockets are required for close to optimum tile stability.

B.S. NAU, BHRA FLUID ENGINEERING, U.K.

Silting occurs when there is no rotation, but not when there is rotation. This suggests it might not be possible to start-up after silting at rest. Is this so?

Authors' reply

When the seal becomes silted the clearance gap closes and start up is not possible if the sealed fluid pressure is maintained. Silting can be cleared rapidly by reducing the sealed pressure to zero and rotating the counterface, after which normal start up is possible.

GENERAL CONTRIBUTIONS TO SESSIONS A, E AND H
A. CAMERON-JOHNSON, SAUER U.K. LTD.

As a potential customer for mechanical seals I have listened with great interest to the speakers at this session. We have heard about research and theoretical analysis which has been performed on face seals with convergence and no vibration for high pressure, seals showing vibration modes and apparently no convergence, seals subjected to external vibration conditions, and a slot fed hydrostatic seal with a leak rate far in excess of anything permitted in the industry in which I work. To the representatives of the fluid sealing industry assembled here for this conference I would like to issue a challenge to produce a cheap, reliable, off-the-shelf mechanical seal to deal with pressures ranging from the top limit of a conventional lip seal up to, say, 10 bar, for shaft sizes in the 25 to 45 mm range, and requiring a housing recess not too much more than the lip seal it might replace.

For reply see page Z70.

Discussion & Contributions

SESSION B : CLEARANCE SEALS

Chairman: Professor J.P. O'Donoghue.
Northern Ireland Polytechnic, U.K.

Papers:

B1   The labyrinth/hydrodynamic disc seal (LHDS).
M.T. Thew, Southampton University, U.K.

B2   Concentric and eccentric running screw seals with laminar flow: Theory and experiment.
K. Heitel, Stuttgart University, Federal Republic of Germany

B3   Compressible labyrinth seal flow with fluctuating upstream pressure.
N. Tully, Natal University, South Africa

Notes:

1.   The author in bold print presented the paper.

2.   Erratum on Paper B3 is provided overleaf.

ERRATUM

PAPER B3

Page B3-41

"CONCULSIONS" should read "CONCLUSIONS"

Authors' replies follow the individual questions.

PAPER B1

The labyrinth/hydrodynamic disc seal (LHDS)

M.T. Thew

Southampton University, U.K.

R.A. GRIMSTON, FLEXIBOX LTD., U.K.

Since the seal described suffers penalties in both length and diameter there seems an opportunity to use these in the form of a conically stepped labyrinth.

This seems to have advantages of better energy destruction at the labyrinth stages and spreads the parasite pressure loss on the upstream side of the disc along the labyrinth stages.

Has this been tried and does it really have the advantages suggested?

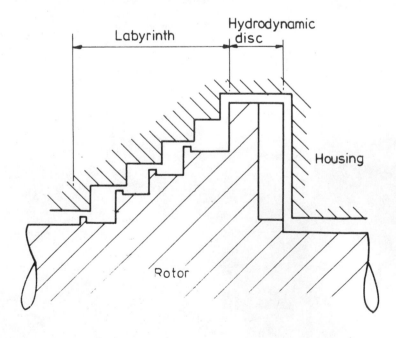

Author's reply

As pointed out, the conical labyrinth is much more effective for energy dissipation than the straight-through labyrinth, as, unlike the latter, it does not suffer with carry-over. The straight-through labyrinth will permit slightly greater axial end float, but the primary reasons for using it were simplicity and the ease with which the geometry may be changed.

Because the rotor of the conical labyrinth is larger in diameter towards the H.D.S. end, it will have a larger power consumption. The effect is not likely to be very significant as most of the power consumption occurs in the H.D.S.

As suggested in the paper, the labyrinth section of the seal could be combined with the H.D.S. Three versions of this are shown on the sketch below and unlike the axial labyrinth, interlocking teeth are easily arranged. Whilst not saving in diameter, the length is now little more than the H.D.S. Again, as compared with the design actually tested, the power consumption of the new versions will be a little higher.

Only the version shown in the paper has actually been tried, but the proposed new variants will work and will be more compact. With more effective energy dissipation, the rather high through flow rates used in the experiments could be considerably reduced.

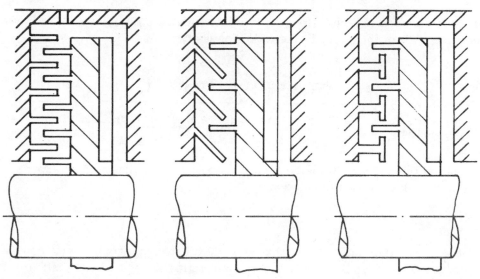

GENERAL CONTRIBUTION TO SESSION B
T.C. CHIVERS, C.E.G.B., U.K.

The authors in this session have considered various forms of clearance seals and have included parameters such as eccentricity; compressible and non-compressible flow. I would like to ask them for their views on clearance seal selection where a shaft is rotating eccentrically in a gaseous medium. Should I use a labyrinth or a screw type?

Author's reply

You pose a rather general question on seal selection if one could apply a single merit parameter analogous to a drag coefficient. The L.H.D.S. is inapplicable to the gaseous medium, which is a slight simplification at least. One would like to have more information on the proposed applications, e.g.:

gas density and viscosity;	shaft surface speed;
size of pressure differential to be sealed;	what leakage is permissible;
	radial and axial end float;
shape of shaft orbit (in the housing);	degree of eccentricity;
	allowable cost of seal.

The labyrinth will tend to have its performance less affected by changes in eccentricity, but the screw type offers more potential for getting smaller leak rates, particularly if the shaft surface speed is high enough to help achieve turbulent flow. The labyrinth may allow materials where a touch would not be catastrophic e.g. the use of a light honeycomb to line the stator.

Concentric and eccentric running screw seals with laminar flow : Theory and experiment.

K. Heitel

Stuttgart University, Federal Republic of Germany

B.S. NAU, BHRA FLUID ENGINEERING, U.K.

The analysis does not seem to include the squeeze film effect, due to $\partial h/\partial t$. Could you clarify this. It would seem that this term would become more important at larger e-values and might therefore explain the observed discrepancy between theory and experiment at larger e-values. It would be possible to write $\partial h/\partial t$ explicitly since h(t) is a simple function.

Author's reply

The introduction of the term $\partial h/\partial t$ or $\dot{h}$ into the basic equations of the Finite-Element -Method changes Eq. (8) of the paper in the following way:-

$$J^{(e)} = \underbrace{\{p\}^T[k]^{(e)}\{p\}}_{\substack{\text{pressure term} \\ \text{with fluidity-} \\ \text{matrix}}} - \underbrace{h^{(e)}A^{(e)}\{V\}^T[B]\{p\}}_{\text{velocity term}} + \underbrace{2\,A^{(e)}[N]^{(e)}\{p\}\dot{h}}_{\substack{\text{squeeze-velocity-} \\ \text{term}}} + \underbrace{2q[N]^{(e)}\{p\}B_2}_{\substack{\text{flow-quantity-} \\ \text{term}}}$$

If you look on the squeeze-velocity-term you may realise that no new expression besides $\dot{h}$ occurs. And $\dot{h}$ is indeed a simple function:

$$\dot{h} = \omega \cdot e \cdot \sin(\omega t)$$

I am not sure whether this formulation may yield better agreement between theory and experiment at larger ϵ-values because on such low rotation speeds of the shaft (n = 100 ÷ 250 rpm) as used in experiments the squeeze-effect could hardly have a high influence on the sealing pressure difference. There is only a discrepancy between theory and experiment concerning the values of the sealing-pressure difference $\Delta\bar{p}_o$. Concerning the pressure distribution in the annulus ($\bar{p}$) there is a very good agreement. And I think if there is generally a high influence of the squeeze-effect there should be also an influence on the pressure distribution.

I think the discrepancy can be explained by the two reasons described in section 5.

I have to thank you for your question and I think further investigations should take into account this discussion and examine the influence of the squeeze-effect in the eccentric running screw-seal on the theoretical results.

P.G. MOLARI, UNIVERSITY OF BOLOGNA, ITALY.

Please can you explain how you have changed this inherently three dimensional problem into a plane one in the sense of what is the field of validity of this assumption?

Author's reply

As I stated in section 2.1, one of the basic assumptions in order to simplify the Navier-Stokes equations is point 3 in that section: The radial clearance h_D and the groove depth t are small in relation to the diameter of the shaft D, so gravity-, inertia- and centrifugal forces are negligible and the viscous forces are dominant. This is only valid in cases when the Reynolds number in the grooves Re_G is much lower than 1 ($Re_G \ll 1$). It should be noted that the Reynolds number has a special definition with screw seals (Please see the answer to Dr. Chivers below).

The above assumptions allow change of the cylindric annulus in to a plane one, as shown in Fig. 2 of the paper.

T.C. CHIVERS, C.E.G.B., U.K.

The authors in this session have considered various forms of clearance seals and have included parameters such as eccentricity, compressible and non-compressible flow. I would like to ask them for their views on clearance seal selection where a shaft is rotating eccentrically in a gaseous medium. Should I use a labyrinth or a screw type?

Author's reply

It is difficult to give an answer on this question because the question is not specified enough. The decision for using a labyrinth or a screw type seal for gas-sealing depends on the following operation-questions:-
1. Is a certain amount of leakage allowed?
2. What is the sealing pressure difference?
3. What is the shaft speed?
4. What happens at zero shaft speed?
5. Is the use of auxiliary pumps possible?
6. Is cooling possible?
   Some remarks on these questions:-
1. If no leakage is allowed, the labyrinth type can't be used. The screw seal can reach zero leakage if a barrier-fluid can be used.
2. The sealing ability of these seal-types depends mainly on the viscosity and therefore on the temperature of the medium.
   With the labyrinth-type you will need a very long seal, if the pressure difference is high and with the screw-type you will need a barrier-fluid.
3. At high shaft speeds ($Re_G > 30$);

$$Re_G = \frac{\varrho \cdot n}{\eta} H^2 \cdot h_D{}^2$$

the screw type is getting turbulent and may be instable. This may also happen at lower shaft speeds if the eccentricity is high. In this case it is better to take a labyrinth-seal. Another possibility with the screw seal is to avoid eccentricity by supporting the sleeve elastic which is possible with 0-ring seals.
4. The main disadvantage of the screw-seal is its inability in sealing at low or zero shaft speeds and therefore secondary sealings are necessary.
5. At high pressure differences or low shaft speeds the screw-seal will need a barrier-fluid which has to be pumped into the annulus. There is also a need for cooling fluid, which has to be pumped between sleeve and housing to get constant viscosity, in the annulus.
6. See point 5.

PAPER B3

Compressible labyrinth seal flow with fluctuating upstream pressure.

N. Tully

Natal University, South Africa.

M.T. THEW, SOUTHAMPTON UNIVERSITY, U.K.

1. What value has been used for γ in the calculations? As some flow machines have shown an effective index for γ differing from the Cp/Cv, could you comment on how the results might change if one used an index other than 1.4?
2. The best form of labyrinth seal - where best may be taken as minimising leakage flow - has been broadly derived from steady flow analysis and experiment. With the results tending to show a fall in leakage as the storage parameter is increased, it may be that the best proportions appropriate to a fluctuating inlet pressure will differ from those for steady flow. Would you please comment?
3. With high pressure gas - hence high density - could enlarged annular cavities such as might be used to increase S, perhaps give rise to Helmholtz resonance conditions and, if so, would this be favourable?

Author's reply

The results presented are for $\gamma = 1.4$. Unfortunately the computer program is not available to check the influence of variable index of expansion, but in general flow is a weak function of the index. For the flow configuration here it is improbable that the acceleration process would be far from isentropic, and for most common gases at low pressures and temperatures γ is close to 1.4.

The principal concern in steady flow labyrinth seals is to achieve adequate energy dissipation between orifices, i.e. to limit carryover of energy. The main configuration parameter concerned is the orifice spacing, which should be minimised to limit the physical length of the seal. For dynamic conditions the spacing may be maintained at the optimum static condition and the pocket volume varied by its depth.

Helmholtz resonance is normally associated with a volume having a single access tube in which it may be assumed that a quantity of gas oscillates as an inertial mass, giving rise to the analogy with a simple spring-mass system. In principal the cavities of a labyrinth could resonate and this would be desirable in raising the phase shift. However in the short orifices used the gas inertia will be very small resulting in an extremely high resonant frequency. Since the fluid velocity is distributed there is no way of making a simple estimate of this frequency.

Discussion & Contributions

SESSION C : LIP SEALS

Chairman: Mr. H.S. Stephens.
BHRA Fluid Engineering, U.K.

Papers:

C1   Using the frictional torque of rotary shaft seals to estimate the film parameters and elastomer surface characteristics.
D.E. Johnston, George Angus & Co. Ltd., U.K.

C2   Effect of water on sealing characteristics of oil seals.
H. Tanoue, M. Ohtaki and H. Hirabayashi, Nippon Oil Seal Industry Co. Ltd., Japan.

C3   The effect of surface roughness and thermal operating conditions on the under-lip temperature of a rotary lip seal.
D.J. Lines, Joseph Lucas Ltd., U.K. and **J.P. O'Donoghue,** Northern Ireland Polytechnic, U.K.

C4   Studies on the radial load of radial shaft lip seals.
F. Ridderskamp, Carl Freudenberg Simrit-Werk, Federal Republic of Germany.

Notes:

1. The author in bold print presented the paper.

2. Paper C2 was presented by M. Kaneta, Kyushu Institute of Technology, Japan.

3. Additional material on Paper C2 is provided overleaf.

ADDITIONAL MATERIAL ON PAPER C2
Effect of water on sealing characteristics of oil seals.
H. Tanoue, M. Ohtaki and H. Hirabayashi,
Nippon Oil Seal Industry Co. Ltd., Japan.

Indications of a dry condition (by dotted line) in Figs. 5 and 8, **Examples of f-G characteristics in the case of sealing the water** (1) and (2), are expressed as a supplement, as follows:-

The coefficient of friction obtained under the dry condition cannot be plotted on the f-G graph. However, each plotted G-value in Figs. 5 and 8 changes in accordance with the change of a shaft speed beacuse the viscosity of water is low and is almost determined by each water temperature (room temperature or 80°C controlled). So, in order to evaluate the coefficient of friction shown in Fig. 3 as an index of the change of the lubricating condition between a seal lip and a shaft, the coefficient of friction has been plotted for each shaft speed under the condition corresponding with each measured value on the f-G graph. Our test results showed that, when the coefficient of friction of oil seals under the water lubricating condition increased to the range of the coefficient under the dry condition, squeaking phenomena and a vapour spouting occurred. And, in order to relate those facts to the condition of an oil film between the seal lip and the shaft, the indication (dotted line) in Figs. 5 and 8 is considered to be worth while.

ERRATA

PAPER C2

Page C2-23

Section 2, Line 11:

"... sealed (12)..." should read "...sealed (13)..."

Page C2-28

Fig. 1(a):
(12) should read (13)

Fig. 1, Explanation of symbols:-
(13) CITY WATER should read:-
(13) LIQUID TO BE SEALED

Authors' replies follow the individual questions.

PAPER C1

Using the functional torque of rotary shaft seals to estimate the film parameters and elastomer surface characteristics.

D.E. Johnston

George Angus & Co. Ltd., U.K.

P.D. SWALES, UNIVERSITY OF LEEDS, U.K.

At first sight I am unhappy about the implication that in a purely hydrodynamic situation we can have a residual torque, at zero velocity. Looking at the equations it seems that Eq. 5 contains an implied V/v term which is indeterminate at zero velocity, and which would account for the discrepancy. For real seals the torque is finite at zero velocity because the rubber asperities are in contact with those of the shaft. Perhaps you would clarify this situation.

Author's reply

There is in fact no implied V/v term in Eq. 5, which is obtained by subtracting Eq. 2 (multiplied by a numerical constant) from Eq. 4. In the mathematical sense the finite torque at zero speed arises from the requirement that the film pressure has to always support the constant lip load. From Eq. 2, it implies that as the speed tends to zero the value of k tends to infinity, which in turn means the minimum film thickness is going infinitely small. Naturally there will be a practical limit of μν below which there will be some solid contact between the seal and the shaft, but provided the experimental measurements are taken in the hydrodynamic region, there is no mathematical inconsistency in extrapolating to zero speed, assuming hypothetical hydrodynamic action.

G.W. HALLIDAY, GEORGE ANGUS & CO. LTD., U.K.

In reply to a comment by Dr. Swales he should bear in mind the existance of visco-elastic properties of rubber. This enables the rubber to behave elastically at very high frequencies and thus react to the asperities at high peripheral speeds.

Author's reply

The surface roughness measurements were made axially across the shaft and I would agree that the circumferential trace is more relevant.

P.H.H. G OUGH, BRITISH LEYLAND CARS LTD., U.K.

While I am sure that you have experience of the importance of the shaft surface texture on seal bedding and life, I would like to hear your views on the texture requirements relative to the maintenance of the oil film adjacent to the seal lip.
The bedding action of the seal strops the shaft so smooth that start-up lubrication, for a splash seal, relies on the meniscus formed either side of the lip.
From a seal user's point of view I feel that torque loss is very closely allied to seal bedding requirements.
Would your work on this aspect form the basis of another paper?

Author's reply

There are three characteristically different modes in which a seal can operate. As a new seal it has to bed-in to form a stable bearing surface, which in the process modifies the surface finish on the shaft. Then, as we know, it runs on a thin film of fluid when the shaft is rotating in the presence of oil, and finally there are the subsequent starts where there is normally no detectable change in the amount of seal wear.
For the maintenance of an oil film it is necessary that there are irregularities on the surface of the seal or the shaft. Their shape and size will govern the torque characteristics. Once the oil is there, surface tension forces will retain it but when the shaft stops, the hydrodynamic lift ceases and the oil will, in the main, be squeezed out.
When shaft rotation recommences a very small quantity of oil in one pocket (either in the shaft or seal) will be sufficient to create some hydrodynamic lift, surface tension will then pull more fluid into the gap thus formed, until a stable state is reached. Extremely small quantities of oil are required by the film that even the residue clinging to the shaft and adjacent the lip will be surfice to provide lubrication immediately on start-up. Thereafter it is possible that axial movement of the fluid in the film coupled with centrifugal forces would ultimately cause a loss with attendant dry running. Even splash lubrication would counteract this.
As regard your last point, it is our intention in the future to look at different shaft surface finishes to see the effects on the seal torque, and hence the characteristics of the bedded-in surface.

T.C. CHIVERS, C.E.G.B., U.K.

The papers on radial lip seals make reference to surface roughness values in terms of R_a alone. This parameter gives information on height distribution only, and no reference to longitudinal aspects is made, yet asperity angle is of relevance to both the hydrodynamics and surface deflection. However, there appears to be an empirical correlation between R_a and the standard deviation of surface slope. I must at this stage confess a general ignorance of lip seal theory but it would appear that the wedge angles quoted in Paper C1 are much greater than those expected on the "optimum" shaft finishes of Paper C3; a simple view suggests that the elastomer features would dominate. It is also known that in operation both the elastomer and shaft surfaces are modified by wear and it is these features that dictate performance and not the as manufactured state.

Bearing in mind my general lack of knowledge of this subject I would welcome your comments on my observations.

Author's reply

The use of a single parameter to define surface roughness, e.g. R_a, is not sufficient data upon which to base a mathematical analysis. However, by stating the type of process used for the finishing operation, a quantitative appreciation can be made, and if the same process is used to produce different roughnesses for the experiments, the comparisons are more meaningful than if different ones are used.

In the experiments described in my paper all the shafts were ground to the same finish, within $\pm$ 10%, using the same grinding wheel, and therefore the comparative measurements between the various elastomers were on a consistent basis. The theory that I have used does not specifically require the surface roughness of the seal to create the hydrodynamic action, although the wedge shape is necessary, but the angles that evolved from the calculations suggested that the seal surface was predominant.

Also, as you point out, the effect of wear must be taken into account, but the influence of the initial conditions cannot be totally ignored. The fact that the shaft seal track can undergo a substantial change in surface finish during operation is another reason for my assigning the hydrodynamic action to the seal surface. The rubber surface is, in my opinion, less likely to exhibit such a dramatic change as wear takes place. Since the elastomer mix comprises filler particles in a matrix, abrading the surface will merely dislodge some of the particles and present a fresh surface of a similar texture.

B.S. NAU, BHRA FLUID ENGINEERING, U.K.

In your paper you refer to rubber asperities on the one hand and shaft asperities on the other. Stating that when run-in the shaft has 'no detectable roughness', I presume this means circumferentially and wonder if in your reply you could present a circumferential Talysurf trace of a wear track. This is however only part of the story since there is also the possibility of shaft out-of-rounds as illustrated by the General Motors workers. This out-of-round would not wear in as readily and might therefore be important in conjunction with rubber hysteresis in generating hydrodynamic lubrication. Professor Hirano has of course put forward a model based on such visco-elastic behaviour. Perhaps one cannot disregard either the rubber or the shaft surface profiles.

Author's reply

All the surface roughness measurements taken during the experiments were axially along the shaft and not circumferentially as would be more apt. However, the lack of the appropriate equipment dictated the course taken. On this point I am thus unable to supply a circumferential Talysurf trace and for this reply the time limit prevents me from obtaining the trace from elsewhere. However, as a general observation if the axial trace is taken at a number of places on the shaft surface before and after testing a seal, and the seal track shows the peaks to have been worn off, it is highly likely that a circumferential trace will exhibit similar characteristics since the peaks should be of a similar height circumferentially (although not necessarily of the same base length measured axially).

In the theory it was not initially assumed the shaft surface roughness had no influence, it so happened that the semi-empirical values for the asperity profile angles appeared too high to be associated with the shaft, when the results were analysed. In addition it is well-known that the shaft is worn smoother under the lip but there is no apparent change in the sealing mechanism therefore it was deduced that the seal surface roughness played the major role.

The other aspect you mention of shaft out-of-roundness is of interest, but as I am not in possession of the literature I cannot give any detailed comment. However, I can visualise a situation where an effective non-uniform shaft radius would cause a hydrodynamic lift similar to that in a journal bearing. If it is assumed that the seal lip does not follow exactly the radial movement of the shaft surface, but that any wedge angle thus created between them is small the coefficient of dynamic friction if proportional to h/l where h is the film thickness and l the length of the wedge. Furthermore, if l is related to the distance between the peaks of any eccentric movement – and these, I presume, are on the macroscopic scale – then using experimentally measured values of the friction coefficient would put l at a similar order of magnitude as h, and this comes out in Professor Hirano's work.

I do not completely discount the possibility that both the surface roughness of the shaft and the seal influence the hydrodynamic film but in this case the changing surface roughness of the shaft, with time, would become an influential factor.

S.J. EDWARDS, PIONIER LAURA BV, NETHERLANDS.

What is the basis for the introduction of split shaft radial load measuring devices as a firm definitive method of measurement which to date has been regarded as a useful comparator but not definitive?

Would you state your reasons for this apparent change of heart? How does the split shaft method compare with other methods of measurement, i.e. incremental split shafts (Williams Lab.)?

One further comment: Split shafts methods for quality control are very acceptable but more work must be carried out and published before one can use the measured values from such machines before seal theory can be based on the values obtained from unproved machines for such definitive purposes.

Author's reply

The theory I have used in my paper assumes that the unit radial load of a seal is constant, and in the main an attempt has been made to compare the torque characteristics of different elastomers on this basis, provided the radial load measuring equipment is consistent in its readings, meaningful comparisons can be made.

Naturally one wishes that the measurements are exact but a certain amount of error is to be expected and indeed has to be tolerated. With the split mandrel technique of measuring radial load this error can be kept small by careful and regular calibration. Admittedly this equipment gives a summation of the radial forces around the circumference and the reading is affected by friction between the seal contact edge and the mandrel. However, the former aspect was all that was required by the theory and the latter was less significant than normal because the seals were measured immediately after test and were therefore covered with a film of hot oil.

As regards the Williams Laboratory equipment you specifically mention I have no first-hand experience of it. From what literature I have seen, it measures the local force over a very small arc through a separated piece in the shaft connected to a force transducer. The fact that there is a separate piece means that there could be edge effects where the rubber deforms around the segment, and hence the integrated value which gives the total radial load will be susceptible to error. This equipment would be of use in detecting lip flaws. On the whole for the measurement I was wanting, the simple split mandrel rig gave me the required accuracy.

PAPER C2

Effect of water on sealing characteristics of oil seals.

H. Tanoue, M. Ohtaki and H. Hirabayashi

Nippon Oil Seal Industry Co. Ltd., Japan

D. HUHN, GUSTAV HUHN AB, SWEDEN.

What material was the shaft on which the test seal was run?

Authors' reply

As the explanation on shafts used in the testing machines shown in Fig. 1 (a) and (b) is insufficient, we supplement shaft materials, etc., as follows:-

Testing machine	Fig. 1 (a)	Fig. 1 (b)
Shaft material	19Cr-9Ni austenite stainless steel (AISI 304) Composition: 0.04% C, 0.74% Si, 1.56% Mn, 19.4% Cr, 9.2% Ni.	
Machining	Plunge grinding	Grinding
Surface roughness	0.8 to 1.0 μm R max	0.8 to 1.0 μm R max.

PAPER C3

The effect of surface roughness and thermal operating conditions on the under-lip temperature of a rotary lip seal.

D.J. Lines

Joseph Lucas Ltd., U.K.

and J.P. O'Donoghue

Northern Ireland Polytechnic, U.K.

G.W. HALLIDAY, GEORGE ANGUS & CO., U.K.

Were the surface roughness measurements made axially across the shaft? The circumferential trace is surely more relevant. Also I agree with Dr. Johnston that no matter what the initial CLA value is after extended running the wear profile is much smoother.

Authors' reply

The roughness was measured axially and circumferentially. The surfaces for the results presented were not "run-in". In other work previously published, the worn surface of the shaft was characterised by smooth peaks while the valleys remained in the original state.

P.D. SWALES, UNIVERSITY OF LEEDS, U.K.

Under rotating conditions at seal speeds of 20 m/s for the rubber to drape itself over successive seal asperities would require the surface to vibrate at 10^5 to 10^6 Hz. Does it not seem likely that the rubber could not accommodate this frequency and that lift will be generated by both the rubber and shaft asperities?

Authors' reply

It is essential to develop a realistic model of what is happening in the contact zone of a seal. The average pressure caused by the interference of the lip is of the order 5 atmospheres, and whatever else happens in the film, this parting force must be generated. The elasticity of the rubber is such that local pressures of the order of 20-50 atmospheres will certainly cause major deformations and the existance of rubber asperities carrying these peak loads as stationary loads is quite beyond the bounds of practical reality. This is not the case for the rigid material. The rubber asperities may have a part to play but only in terms of rate of deformation over a short period in contact with an asperity of the rigid shaft. This is where the visco-elastic performance is likely to be of major importance.

The only experimental case for the existance of a predominant visco-elastic effect was proposed by Hirano to explain his observed speed effect, although he noted with concern that temperature changes the visco-elastic properties in such a manner as to reverse the required trend. In considering the heat balance and underlip temperature we were able to fully explain the Japanese results without including the visco-elastic effect.

S.J. EDWARDS, PIONIER LAURA BV, NETHERLANDS.

What is the basis for the introduction of split shaft radial load measuring devices as a firm definitive method of measurement which to date has been regarded as a useful comparator but not definitive?

Would the authors state their reasons for this apparent change of heart? How does the split shaft method compare with other methods of measurement, i.e. incremental split shafts (Williams Lab.)?

One further comment: Split shafts methods for quality control are very acceptable but more work must be carried out and published before one can use the measured values from such machines before seal theory can be based on the values obtained from unproved machines for such definitive purposes.

Authors' reply

There is no change of heart!

We have never used any other method, and perhaps now the question has been raised it is of interest to the authors to look at other methods of measuring radial load. At no time have we suggested that the split shaft is definitive, but it has proved a repeatable and useful tool in our work on lip seals.

B.S. NAU, BHRA FLUID ENGINEERING, U.K.

Your presentation showed a rubber seal "draped" over an asperity like a mountain range. In reality a prairie would be nearer the truth. Would you elaborate on the implications of this aspect.

Authors' reply

It is always difficult to defend the grotesque distortion of scale necessary to explain the action of a hydrodynamic wedge. Since 1886 when Reynolds first drew his classic explanation of continuity of flow in a converging film, it has been customary to picture the mechanism using a variation of film thickness of the same order as the length of the bearing! Could you present a scale picture of a film of the order of microns and a surface of the order of centimetres you would indeed do a service to truth, but could you then explain the mechanism of lubrication?

In teaching journal bearing lubrication we commonly use clearances 30% of the shaft diameter to illustrate the eccentric film formation, but students do not assume that this is a typical design figure. In fact the theory demands that the surface should be more like a prairie than the Andes since this is an assumption in Reynolds analysis. The surface variations in height must be small compared to the rubber seal dimensions for EHL theory to apply.

T.C. CHIVERS, C.E.G.B., U.K.

The papers on radial lip seals make reference to surface roughness values in terms of R_a alone. This parameter gives information on height distribution only, and no reference to longitudinal aspects is made, yet asperity angle is of relevance to both the hydrodynamics and surface deflection. However, there appears to be an empirical correlation between R_a and the standard deviation of surface slope. I must at this stage confess a general ignorance of lip seal theory but it would appear that the wedge angles quoted in Paper C1 are much greater than those expected on the "optimum" shaft finishes of Paper C3; a simple view suggests that the elastomer features would dominate. It is also known that in operation both the elastomer and shaft surfaces are modified by wear and it is these features that dictate performance and not the as manufactured state.

Bearing in mind my general lack of knowledge of this subject I would welcome the authors comments on my observations.

Authors' reply

The surfaces in the experiments referred to in our paper were not "run-in" and consequently the relationship of R_a to surface shape was a constant, the surfaces being plunge ground to give a roughness with no directional characteristics.

I do not think that a "wedge" gives a true picture of the lubrication condition. Very thin films exist over elastohydrodynamic contacts on the rigid asperities, these areas correspond to the static interference contact areas of the shaft and seal with similar pressures due to the contact pressures deforming the rubber.

D. HUHN, GUSTAV HUHN AB, SWEDEN.

1. Referring to Fig. 8 of the paper, could you please explain what material is "Metco 18C"? This figure shows clearly the importance of the right shaft material with respect to thermal conductivity. Have results been obtained with bronze or for instance containing 90% Cu and 10% Sn? This material would probably give an even lower underlip temperature.

2. As can be seen from Fig. 4 in the paper, the contact area of a lip seal with the shaft is very restricted and therefore the cooling of the contact area becomes more important than with other types of seals - mechanical seals and circumferential seals. Experience gained with circumferential seals of larger dimensions and reported in Paper E3 at the 6th ICFS in Munich (1973), shows that inspite of their inferior strength values of bronze in comparison with stainless steel, less wear in clean water and in clean oil has been obtained with bronze as shaft liner material due to its better heat dissipation.

Authors' reply

Metco 18C is a hard facing nickel alloy material which had the desired property of low conductivity to enable us to compare underlip temperatures with those for steel and brass shafts.

We would agree that with a 90% Cu and 10% Sn bronze would give lower temperatures and that the restricted heat flow into the shaft results in temperatures approximately inverse to the conductivity coefficient, and that from Hirano's Theory this would lead to higher film thicknesses, lower wear and at the same time provide an effective seal.

PAPER C4

Studies on the radial load of radial shaft lip seals.

F. Ridderskamp

Carl Freudenberg Simrit-Werk, Federal Republic of Germany

M.W. ASTON, LUCAS AEROSPACE, U.K.

In measuring the radial load by method 2 the stress relaxation characteristics of the rubber are being examined in the diametral mode. Are the results shown based on measurements made at room temperature or at the aging temperature?

Author's reply

All radial load measurements are made:
a.   before aging: at room temperature (new oil seals; that means $23^{\circ}C \pm 5^{\circ}$)
b.   during aging: at "aging temperature" (NBR $100^{\circ}C$, ACM and FKM $130^{\circ}C$).
That means the results are based on measurements made at room temperature.

S.J. EDWARDS, PIONIER LAURA BV, NETHERLANDS.

What is the basis for the introduction of split shaft radial load measuring devices as a firm definitive method of measurement which to date has been regarded as a useful comparator but not definitive?

Would you state your reasons for this apparent change of heart? How does the split shaft method compare with other methods of measurement, i.e. incremental split shafts (Williams Lab.)?

One further comment: Split shaft methods for quality control are very acceptable but more work must be carried out and published before one can use the measured values from such machines before seal theory can be based on the values obtained from unproved machines for such definitive purposes.

Author's reply

Radial-load-measuring with a split mandrel measuring device is the standard measuring method all over Eruope (especially in the car- and truck- industries). Two different systems are used for measuring:-
a.   The 'Angus-Radial-Load-Rig', and
b.   The 'Freudenberg Radiameters'.
Both methods have appeared during about the last ten years:
1.   Acceptable for production control.
2.   Acceptable for development and engineering.
3.   Acceptable for inspection in the automotive industries.
The measured values between the Angus and Freudenberg methods are comparable. There was a difference of about 3%. This difference is caused by the fact that there is a small change in shaft-diameter for the Freudenberg-split mandrel during the measuring

operation, while the diameter does not change in the Angus split mandrel.

A study on the Angus and Freudenberg methods was made in 1970 and can be published if anybody has an interest in this report.

An investigation between split-mandrel-devices and a measuring device from Williams Lab. does not exist in our company.

Discussion & Contributions

Notes:

1.   The author in bold print presented the paper.

2.   Errata on Paper D1 is provided overleaf.

ERRATA

PAPER D1

   Page D1-6

   Line 11 should read:-

   144 x 118 x 5

Page D1-7

   Fig. 2, at top, should read:-

   4" ASA 600 lbs class spiral wound
   gaskets. Flange finish 3, 125 μm Ra.

   Fig. 3, at top, should read:

   4" ASA 600 lbs class spiral wound
   gaskets. Flange finish 3, 125 μm Ra.

Page D1-8

   Fig. 5, at top, should read:

   4" ASA 600 lbs class spiral wound
   gaskets. Flange finish 3, 125 μm Ra.

Page D1-9

   Fig. 6, at bottom, should read:

   Leakage pressure v gasket loading graph. Gasket
   of expanded PTFE $\emptyset$ 6, 4 Flange finish - 6, 5 μm Ra.

Authors' replies follow the individual questions.

PAPER D1

A review of the properties of expanded graphite and selected forms of PTFE as alternatives to asbestos in a sealing role.

E. Staaf

Swedish State Power Board, Sweden.

J. ABBOTT, FLEXITALLIC GASKETS LTD., U.K.

Graphite filled spiral wound gaskets clearly show a superior sealing performance to asbestos filled gaskets, but the results of the comparison tests indicated in Fig. 1 give a misleading indication of the performance of asbestos paper filled gaskets. One of the benefits of asbestos paper is that the gasket density can be matched to the bolt load and operating pressure. It is therefore possible to produce a high density asbestos paper filled gasket which will give a sealing performance approaching that of a graphite filled spiral wound gasket, when operating on a flange finish of 3.2 micro metres. The graphite filled gasket has the added advantage of being capable of sealing on standard BS 1560 flanges having a surface finish of 10 micro metres even under hazardous gas duties.

Author's reply

The asbestos-paper filled gasket in Fig. 1 was not produced by Flexitallic. Tests were made later with two off 4"/600 lbs ASA spiral wound gaskets manufactured by Flexitallic. The test results are given in the figure overleaf.

The leakage curve for "Flexitallic 1" has approximately the same form as the criticized curve in Fig. 1 whilst the curve for "Flexitallic 2" supports your opinion.

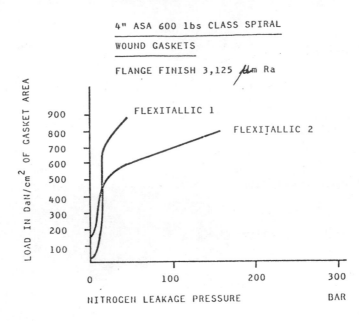

4" ASA 600 lbs CLASS SPIRAL WOUND GASKETS

FLANGE FINISH 3,125 μm Ra

N.E. JOHNSON, SEALOL INC., U.S.A.

Most of your paper is devoted to static packing testing on expanded graphite. We are testing expanded graphite in dynamic high pressure, high temperature applications for nuclear service and I wonder if you would care to comment on any dynamic testing you have done on this material.

Author's reply

No dynamic tests have been carried out.

P. TAYLOR, MINISTRY OF DEFENCE, U.K.

On the risk of corrosion from the use of expanded graphite I note that although much alarm has been expressed about the possible problem due to galvanic action between graphite and a metal in contact with it in the presence of an electrolyte you have been able to elicit only two cases of slight corrosion despite wide enquiries. Does this mean that the fears expressed are unfounded or that the evidence is inconclusive and that special controlled tests on this particular aspect should be made?

Author's reply

As the graphite material has shown to have excellent sealing properties over a very long period there has been no need to demount the valves and inspect the valve stems. This does not mean, however, that these fears are without grounds. Therefore special control tests should be carried out to define this problem.

G.L. DOUBT, ATOMIC ENERGY OF CANADA, CANADA.

I would like to offer some information on the risk of corrosion with expanded graphite valve packings that may help explain why there is concern, as mentioned on page 5 of your paper. Tables 1 and 2 are a collection of results from tests done here or under AECL sponsorship over the past 8 years, all at roughly equivalent water conditions (270-300°C, 7-10 MPa, pH10, oxygen concentration less than 100 ppb in tests 1-6, unknown in tests 7-11). In summary; (a) some stem pitting has usually occurred when we have used this material, (b) stem pitting has never been correlated with leakage, (c) leakage has usually been very low relative to that of asbestos packings at comparable gland pressures, (d) we have been alarmed at the tendency for type 30d stainless steel collection lines to corrode through when collecting leakage from packings or gaskets of expanded graphite. Unfortunately our attempts to isolate the cause of this corrosion have been unsuccessful.

The following conditions are common to all tests in the following Tables: (a) All packing rings were preformed in either complete or split form with graphite laminations perpendicular to the axis. (b) All packing glands were double packed with a lantern ring separating primary and secondary packing. Primary packings contained 3 to 6 rings, secondary packings 1 or 2. (c) Leakage was collected from the inter-packing space, at atmospheric pressure, via a small condenser. (d) Excepting test 3 all gland followers were spring loaded to maintain gland pressure. (e) All stuffing boxes were kept dry prior to high temperature operation.

Expanded graphite packings definitely offer better sealing in the above conditions than most woven asbestos packings. However, in my opinion, it should not be used with stainless steel stems for critical applications until it is proven that stem corrosion is self limiting in the long term, or can be avoided by controlling trace impurities in the packing or by strategic choice of materials for other stuffing box components.

Table 1   Leakage and Stem Corrosion in Tests of Expanded Graphite Packing

Test No.	Stem Material (Stnls. Stl. type)	Test Dura- tion (days)	No. of Stem Actua- tions	No. of Cooldown -Reheat Cycles	Total Accum. Leakage (g)	Stem Surface Condition after Test	Notes
1	316	8	44000	2	317	slight pitting in vicinity of inner packing ring	(a) failure by excessive leakage during 2nd cooldown (b) accum. leakage is total before failure (c) expanded graphite was used in a graphite/ reinforced TFE/graphite sandwich
2	410 or 416	660	68	37	900	slight pitting near outer edge of primary packing	(a) failure by excessive leakage during 37th cooldown (b) accum. leakage is total before failure (c) 3 mm diameter, type 304 stainless steel leak collection line[1] corroded through at roughly 3-4 week intervals
3	410 or 416	1500	202	96	34835	region % of area pitted primary packing 1-2 secondary packing 90-95 inter packing space 30-40 Surface in primary packing region was blackened to about 0.01 mm deep and harder than original	(a) valve isolated inlet and outlet for about 5 years between test and inspection (b) corrosion of leak collection line as in test 2[1]
4 (3 replicates)	410	98	70	30	under 2 (each)	region % of area pitted primary packing 0.012 secondary packing 0.32	parallel tests with as- bestos[2] packings yielded ½ and 1/20 as much corro- sion in primary and sec- ondary packing regions respectively

Test No.	Stem Material (stnls. stl. type)	Test Duration (days)	No. of Stem Actuations	No. of Cooldown-Reheat Cycles	Total Accum. Leakage (g)	Stem Surface Condition after Test	Notes
5 (3 replicates)	410	98	70	30	under 2 (each)	region / % of area pitted / primary packing 0.025 / secondary packing 4.2	(a) these valves underwent 6 strong decontamination treatments[3] (b) parallel tests with asbestos[2] packings yielded 1/10 and 1/700 as much corrosion in primary and secondary packing regions respectively
6 (2 replicates)		98	70	30	under 2 (each)	region / % of area pitted / primary packing 19 / secondary packing 6.9	(a) these valves underwent 6 strong decontamination treatments[3] (b) parallel tests with asbestos[2] packings yielded 1/10000 and 1/2000 as much corrosion in primary and secondary packing regions respectively
7 (3 test assemblies)	410	35	25	0	11 (average)	pitting at each end of secondary packing in all cases, at outer edge of primary packing in 2 cases	(a) zinc washers were interleaved with graphite rings in test 9 and 11 (b) parallel tests with asbestos[1] packings were run for comparison with tests 7 to 11. These yielded corrosion in the vicinity of secondary packings in about the same proportion of cases but corrosion was less severe than with graphite packings
8 (3 replicates)	410	14	5000	0	1009 (average)	no corrosion	
9 (2 replicates)		35	5000	0	4376 (average)	slight pitting in region of secondary packing in 1 of the 2 replicates	
10 (2 replicates)		35	5000	0	784 (average)	pitting at each end of secondary packing in 1 of the 2 replicates	
11		35	5000	0	564	no corrosion	

(1) Corrosion of type 304 stainless steel collection lines occurred in a region of steep temperature gradient near the stuffing box. It has also occurred in tests of expanded graphite gaskets.

(2) Zinc impregated, graphite lubricated, inconel reinforced, woven chrysotile asbestos packing.

(3) Alternating 7% base and 2.5% acid solutions in the order base/acid/base/rinse. Acid or base conditions exist for about 4 h during each decontamination.

Reference: Tests 7-11, AECL 6183 Valve Stem Packing Seal Test Results for Primary Heat Transport System Conditions in Canadian Nuclear Generating Stations, D.F. Dixon, J.M. Farrell, R.F. Coutinho.

Tests 1-6, Unpublished test results.

Table 2   Stuffing Box Details

Test No.	Materials (Stainless Steel, AISI type no.)				Diameters		Gland Pressure	Stroke	No. of Test Assemblies	Stem Heat Treatment
	Stem	Stuffing Box	Gland Follower	Lantern Ring	Stem (mm)	Stuffing Box (mm)	(MPa)	(mm)		
1	316	C.S.	316	316	19	35	13	17 (non-rotating)	1	—
2	410 or 416	C.S.	C.S.	416	16	29	42	25 (non-rotating)	1	unknown
3	410 or 416	C.S.	C.S.	416	16	29	23	25 (non-rotating)	1	unknown
4	410	C.S.	C.S.	C.S.	6	13	55	4.5 (3 turns)	3	hardened, tempered at 300°C
5	410	C.S.	C.S.	C.S.	6	13	55	4.5 (3 turns)	3	
6	410	C.S.	C.S.	C.S.	16	29	55	10 (3 turns)	2	
7	410	316	C.S.	304	32	67	35	38 (7.5 turns)	3	1 annealed / 1 hard., temp. at 370°C / 1 hard., temp. at 760°C
8	410	316	C.S.	304	32	67	35	38 (7.5 turns)	3	annealed
9	410	316	C.S.	304	32	67	35	38 (7.5 turns)	2	hard., temp. at 760°C
10	410	316	C.S.	304	32	67	35	38 (7.5 turns)	2	hardened, tempered at 370°C
11	410	316	C.S.	304	32	67	35	38 (7.5 turns)	1	

C.S. = carbon steel

Author's reply

Your report concerning the corrosion effects of graphite material on valve stems is a very interesting supplement to the work carried out with expanded graphite. Furthermore the leakage results support the value of this material.

I agree that one should control the impurities in the material and furthermore select suitable stem materials.

PAPER D3

The status of magnetic liquid seals.

R.L. Bailey

Oxford University, U.K.

J.D. KIBBLE, NATIONAL COAL BOARD, U.K.

It would be interesting to have more information about the ability of magnetic seals to withstand axial movements. Could they be used as reciprocating seals?

Author's reply

Magnetic liquid seals can be made to perform with reasonable success only in sealing axially mobile shafts. They are limited to strokes of less than two millimetres, say, or to slow axial velocities. With long strokes the fluid becomes spread along the shaft in a thin layer and may need to be collected with wipers. In this sealing mode the technique is nowhere as elegant as it is in the rotary mode.

D. REDDY, BHRA FLUID ENGINEERING, U.K.

Is it not in principle possible to use a ferrofluid seal for reciprocating applications, if the length of seal (i.e. No. of stages) is much greater than the length of stroke of the reciprocating shaft? (The reasoning behind the question is that the volume of ferrofluid in the seal would be large in relation to the volume of film being dragged in and out of the seals).

Author's reply

If the volume of ferrofluid is large compared with the volume left behind on the wetted area of the shaft this will help. Note that the liquid will creep back to the seal even in a fairly low field gradient. This means that if the seal is left static for several hours between strokes some of the liquid will creep back to the seal.

R.T. ROWLES, BHRA FLUID ENGINEERING, U.K.

In Fig. 4 you quote a formula for viscous power loss and give a typical curve. Have you obtained experimental correlation for the formula given?

Author's reply

This formula is an adaptation of Petrov's equation for viscous loss in oil films in bearings. It has been shown to be accurate within 20% in this application. The version given in Fig. 4 is a generalised version suitable for an initial estimate of power loss.

Discussion & Contributions

SESSION E : MECHANICAL FACE SEALS II

Chairman: Mr. F. Gibson,
James Walker & Co. Ltd., U.K.

Papers:

E1   Quality control and its place in shaft sealing.
J. Phillips and **C.M. Johnson**, Flexibox Ltd., U.K.

E2   Comparative testing of a number of radial face seals.
A.C. Pijcke, Netherlands Maritime Institute, and **P. de Vries**, Royal
Netherlands Naval College, The Netherlands.

E3   Mechanical seals for aqueous media subject to high pressures.
W. Schopplein, Feodor Burgmann Dichtungswerk, Federal Republic of Germany.

Notes:

1. The author in bold print presented the paper.

2. Erratum on Paper E2 is provided overleaf.

PAPER E2

      Page E2-25

    Fig. 3 should be replaced by the following:

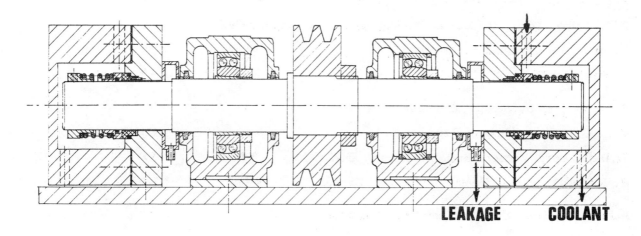

Fig. 3   Seal testing chambers

Authors' replies follow the individual questions.

PAPER E1

Quality control and its place in shaft sealing.

J. Phillips and C.M. Johnson

Flexibox Ltd., U.K.

P.C. BARNARD, ESSO PETROLEUM CO., U.K.

Applying simple logic from Paper C1 to Papers C2 and C3 - if you don't have it, it can't go wrong - i.e. if your problems are from a flush remove it then select face materials that can stand the thermal strain - along with no leakage and you have success. The skill comes in getting the face geometry and material right.

Authors' reply

In an ideal world, the simple logic of "if you don't have it, it can't go wrong" is very true. Unfortunately, quality control has to operate within the limitations of the state of the art, as well as commercial considerations. For example, the geometry of an ideal seal ring may be so complex that it is un-machinable, especially if an exotic material has to be selected from thermal strain considerations. Even if the desired shape of ring can be purchased, the cost may be prohibitive for the particular application. Engineering is often a question of compromise, and compromise might lead us to a flushed seal selection. Let us also remember that the application of flushing to seal faces gives an extremely effective system, and rarely gives rise to problems if properly applied.

C.R. MIANNAY, EXXON CO., U.S.A.

With reference to the quality circle, how does Flexibox rationalise the high amount of spare part replacement by customers for seals that do not operate leak free? Can the replacement of parts indicate the quality circle has not been used or that some section in the quality circle is inadequate.

Authors' reply

The fact that spare seal parts are supplied because of leakage problems does not, in itself, indicate either a breakdown or inadequacy in the sealmaker's part of the quality circle. Poor installation, pump vibrations and changes in the product can all manifest themselves as seal leakage, and these all fall within that part of the quality circle controlled by the user. The sealmaker has little or no control over such causes of leakage but can, and often does, offer excellent training courses for the process industry engineering staff. Our experience has shown that such courses help to increase the ensuing seal life.

Additionally, a number of steps should be taken in resolving any persistent sealing problem viz:-
1. The user must make known the problem to the seal manufacturer.
2. The user and seal manufacturer together must review the specification against which the seal was supplied. The original specification may have been incorrect, or may subsequently have changed.
3. In the light of the revised specification, the selection of the sealing system should be reviewed.
4. The installation of the sealing system should be checked to ensure it conforms to the Supplier's recommendations.

All these checks are implied by the quality circle philosophy and, if pursued, should result in the user getting what he wants.

R. METCALFE, ATOMIC ENERGY OF CANADA, CANADA.

I am happy to see quality being stressed. The problem from the nuclear point of view is that adequate quality control be bought for our most critical applications. Why is this?
1. "Users" would be willing to pay for **guaranteed** quality (Nuclear primary pump seal failures now cost our electrical utility £100,000 each in lost generation), but experience has shown little relationship between cost and quality.
2. On the other side "manufacturers" will not give sufficient guarantee - a guarantee that a replacement seal of the same type will be provided is the last thing the user wants.

Can you comment on how higher quality seals at commensurately higher initial cost can be achieved, because in nuclear service we are definitely not interested in the spare parts business.

Authors' reply

Optimum seal quality is achieved by following the quality control philosophy outlined in the paper. In the nuclear field, great importance must be attached to liaison between customer and supplier in drawing up the precise specification the seal is to meet. This collaboration should also extend into the design stage, as most seal manufacturers are unlikely to be fully aware of all the problems associated with nuclear environments.

In designing seals for some applications in the nuclear industry, the designer is almost certain to be working at the limits of the known state of the art. To produce a seal of guaranteed quality, and this presumably means performance, it will almost certainly be necessary to test the seal in the operating environment.

The answer to the question then is **co-operation**. This should include the seal designer and the end user, and we would be delighted to have the opportunity to participate in such an exercise.

A.C. PIJCKE, NETHERLANDS MARITIME INSTITUTE, THE NETHERLANDS.

Could you give your opinion on the following: Why a separate quality control for the seal and not an integrated Q.C. pump-seal? This seems to be important as you also mention in section 3.6 of your paper (stuffing box geometry).

Authors' reply

As manufacturers of mechanical seals, Flexibox Ltd., can only have responsibility for the quality control of the seal. Of course, quality control of the pump is equally important, and the pump manufacturer needs to have his own quality control procedures. These would include systems for procuring proprietary items, including seals, and hence there is a link between **"PROCUREMENT"** in the pump manufacturer's quality circle and **"SPECIFICATION"** in the seal suppliers quality circle. All relevant information for the seal supplier, including details of the stuffing box geometry, should be conveyed along this link.

Authors' reply to comments by A. Cameron-Johnson
(for comments see Session A, Page   ).

If Mr. Cameron-Johnson will send the authors his seal specification, including a definition of the word "cheap", they will ensure that a proposal is prepared for his consideration.

PAPER E2

Comparative testing of a number of radial face seals.

A.C. Pijcke

Netherlands Maritime Institute

and P. de Vries

Royal Netherlands Naval College

R.A. GRIMSTON, FLEXIBOX LTD., U.K.

Noticing the variation in "Solids Content" of Table IV and bearing in mind complexity of pipe layout, were you able to take any special steps to ensure all seals operated under similar contamination levels (in respect to solids contents, particle size etc.)?

Authors' reply

During the tests the general supply of sea-water was analysed continuously. The sea-water used in each chamber was not analysed continuously, but was checked each two weeks of running. As a result of these checks, we did not find any differences with the analysed composition of the general supply.

P.C. BARNARD, ESSO PETROLEUM CO., U.K.

Were the leakage rates calculated by Poiseuille flow formulation less than or larger than those measured on test?

Authors' reply

In general the measured leakage rates were less than those calculated by laminar Poiseuille formulation.

The differences did vary in the range of 1.1 to 2.25.

F.J. TRIBE, ADMIRALTY MARINE TECHNOLOGY ESTABLISHMENT, U.K.

Could you please state:-
1. How you defined failure?
2. Did you analyse the failures to establish individual mechanisms, particularly with reference to:-
   a. 3 body abrasion of the faces by silt.
   b. Silt preventing compliance of the faces by inhibiting dynamic alignment.
   c. Galvanic corrosion of faces by carbon/graphite.
3. Have you experienced coating failures induced by seawater (or its corroding species) permeating the coatings and attacking the substrates?

Authors' reply

1. We did define a seal failure as follows:-
   - a seal failure occurs, when the leak rate is more than $2.5 \text{ cm}^3.\text{h}^{-1}$ -
2. Before and after a test the seals were checked extensively. All the failures were analysed to establish, if possible, the mechanism of the failure. Regarding the 1st phase testing, we did find that most of the failures were caused by wear of the sealing faces. During the 2nd phase testing period no abrasion of the faces by salt particles or silt was found.

   We did not find inhibiting of the dynamic alignment by silt. Also no galvanic corrosion of the faces was established.

   Parallel to the normal testing programme, we did carry out galvanic corrosion tests. The theoretical conditions for galvanic corrosion are present if 2 different metals are used. In general however the mass of the "positive" metal (for example: shaft material AISI316 is relatively greater than the mass of the "negative" metal (for example: seal material bronze) so the chance for galvanic corrosion to occur will be less than expected. During a test with a seal (mainly bronze), a small hastelloy ball was placed inside the testing chamber. It was found that after some time the ball was covered with traces of bronze material.
3. We did not experience ceramic coating failures, on the contrary. After the test runs all the rings were in an excellent condition.

   Rings, however, with a stellite layer did perform quite badly. Many problems with this type of material were experienced.

D. HUHN, GUSTAV HUHN AB, SWEDEN.

1. Referring to Fig. 3 of the paper, I wonder if the difference in silt being built up in the first seal (Crane) compared with the other (Flexibox) cannot be attributed to the difference in clearance between rotating seal parts and the casing.
2. What materials have been used in shaft and seal casing? Some electrolytic corrosion might be related to the fact that seals made in stainless steel are run in a bronze casing and vice versa.

Authors' reply

1. Fig. 3 shows a general impression of a seal testing unit. As can be seen in Figs. 8, 9, 10, 11b and 12, all the chambers of the testing units do have the exact same dimensions.

   As the Crane and Huhn seals have less clearances between the rotating seal parts and the casing and the fact that these seals did compile silt, we agree with your suggestion that the compiling of silt in these seals could perhaps be explained by the differences in clearance.

2.   We have used for the shaft and seal casing of the testing units, the exact same materials as used in the pumps.

pump-shaft: AISI 316 ; shaft testing unit: AISI 316

pump-casing/impeller: Aluminium Bronze ; casing testing unit: Aluminium bronze.

COMMENT ON PAPER E2:
J. PHILLIPS, FLEXIBOX LTD., U.K.

For the moderate sea water duties quoted the basic Flexibox selection has been, for at least the last four years, solid ceramic versus plain carbon, and it was some of the early tests carried out by Mr. Pijcke that led us to look for this better face combination.

It is also worth noting that not only is ceramic versus carbon a good face pairing for dealing with the abrasive nature of sea water but because ceramic is a good electrical insulator it effectively breaks up some of the circulating current circuits and hence reduces the possibility of galvanic corrosion.

Many thousands of hours of trouble free running have been recorded.

Additionally Flexibox now has very well established techniques for manufacturing large splits seals, up to 165 mm, for cargo/ballast pumps in this face combination.

PAPER E3

Mechanical seals for aqueous media subject to high pressures.

W. Schopplein

Feodor Burgman Dichtungswerk, Federal Republic of Germany.

P.G. MOLARI, UNIVERSITY OF BOLOGNA, ITALY.

How did you choose the boundary conditions on the contact area? In what way did you choose the nodes in contact and the loads in operative (pressures - temperatures) conditions?

Author's reply

In relation to its width the contact area is divided into a number of equal gap-elements. As a first step these elements got an equal load resulting from the hydraulic forces. Later on we tested out the variation using a non linear load distribution.

At the moment our calculations are regarding pressure and temperature deformations as well as the influence of centrifugal forces.

COMMENT ON ABOVE QUESTION BY P.G. MOLARI:
R.A. GRIMSTON, FLEXIBOX LTD., U.K.

The way in which the seal's surroundings are fed into computer are indeed very critical e.g. An 0-ring can be treated as a liquid transferring axial pressure loads freely onto shaft in a radial direction, or as an elastic solid. Both give different results.

Rig confirmation of computer results is very much affected by such problems.

J. DALY, BORG WARNER MECHANICAL SEALS, U.K.

I was interested in the slides showing erosion/corrosion damage to tungsten carbide seal rings on standby pumps in hot water service. Would you comment on the following:-

Seal size, pressures and temperatures being sealed (at standby)? Chemical composition of the tungsten carbide and carbon faces? pH values of the water?

Were investigations made into possible stray currents?

What time scales were involved? The erosion seemed to be concentrated in specific areas notably adjacent to the hydrodynamic slots. Why is this?

How was the problem solved?

Author's reply

Erosion/corrosion problems with mechanical seals may be caused by a great variety of reasons. Usually it is a combination of face distortions; resulting leakage and erosion of the face material. Besides this we know that under certain conditions (e.g. special treated demineralised water) there is a partial corrosion of the face material, weakening up the structure of this material and resulting in leakage. Problems are solved when face deformations are reduced and face materials (SiC) are used that, due to their structure and chemical property, are resistant against the described effects.

R. METCALFE, ATOMIC ENERGY OF CANADA, CANADA.

This was a good presentation of the capabilities of quality seals for high pressure water and nuclear applications. I must comment, however, that the finite element analysis of the flange that was shown was contributed by Atomic Energy of Canada. I therefore wonder whether your company now do such analysis routinely themselves for their flanges and seal components?

Author's reply

It is right that the picture of a finite element analysis plot for a flange was contributed by AEC.

Besides this, Feodor Burgmann, in close cooperation with a computer company, is now able to calculate seal deformations under various conditions using a prepared program.

Since then a lot of work has been done but we are still carrying on to find the best and, due to the high costs, the most efficient way.

Discussion & Contributions

SESSION F : ELASTOMERIC SEALS

Chairman: J.A. Stephens,
Bestobell Seals Ltd., U.K.

Papers:

F1   High temperature elastomers in gas turbine engine fuel sealing.
      K.P. Palmer and **M.W. Aston**, Lucas Aerospace, U.K.

F2   Numerical and experimental stress-strain analysis on rubber-like seals in
      large elastic deformations under unilateral contact.
      G. Medri, P.G. Molari and **A. Strozzi**, Bologna University, Italy.

F3   The achievement of minimum leakage from elastomeric seals.
      T.C. Chivers and **R.P. Hunt**, CEGB Berkeley Nuclear Laboratories, U.K.

Notes:

1. The author in bold print presented the paper.

2. Erratum on Paper F1 is provided overleaf.

ERRATUM

PAPER F1

Page F1–17

Fig. 15, caption should read:-

Fig. 15 Stress relaxation of fluorocarbon '0' rings aged at $200^{\circ}C$ in kerosene (stress measured at room temperature).

Authors' replies follow the individual questions.

PAPER F1

High temperature elastomers in gas turbine engine fuel sealing.

K.P. Palmer and M.W. Aston

Lucas Aerospace, U.K.

G.W. HALLIDAY, GEORGE ANGUS & CO., U.K.

Could you please advise whether the compression set tests referred to were done in accordance with the usual B.S. spec routine which involves 'hot release' of the test specimen.

In our experience a "cold release" technique is more meaningful and more likely to correlate with results of stress relaxation tests.

Authors' reply

Yes. The compression set tests were carried out as BS 903 part A6 which requires hot release of the test piece from compression in the jig.

We would agree that you would get better correlation with compression stress relaxation by cold release but the standard test procedure does not allow for this and you very rarely, if ever, see cold release results stated in manufacturers literature. The British Standard test procedures are aimed at getting good reproducibility and repeatability of results and cold release does not give as good results as hot release and 30 minutes standing on the bench. This is because as soon as the test piece is released it starts to recover particularly on a short term test. It is desirable to reach as near as possible a steady condition and this is better achieved by hot release and a 30 minute stand rather than cold release and immediate measurement.

From our experience there is a greater tendency for the test pieces to stick to the plates with cold release. In fact this is a problem with long term high temperature compression set testing particularly in some fluids and this is one of the many reasons why we prefer compression stress relaxation testing for determining seal performance.

J.A. STEPHENS, BESTOBELL SEALS LTD., U.K.

Although I agree that in most situations Stress Relaxation is the most relevant indicator of sealing performance, I suggest that when panting occurs, that is the deflection/movement of supporting metal parts due to pressure changes, then Compression Set and possibly Rebound Resilience may also be relevant. In these circumstances the seal must recover to follow the movement of the metal part and compression set is a measure of degree of recovery and rebound resilience indicates speed or rate of recovery.

Would you please comment.

Authors' reply

The high pressure which causes the panting also energises the seal and pushes it up into the extrusion gap caused by the pant. Therefore the ability of the rubber to recover in terms of compression set is not relevant to this mode of operation. It is necessary for the seal to have some resilience and not be rock hard in order that it will move under the energizing forces.

Seals can only leak when they have lost their sealing force and are unenergised by pressure which is why our sealing tests incorporate continuous pressure cycling at a rate slow enough to allow the seal to show leakage at a near zero fluid pressure.

H. HOPP, MARTIN MERKEL KG, FEDERAL REPUBLIC OF GERMANY.

Do you have any experience with a 'dynamic-gap' in the test block - Fig. 6? If yes, please could you give us the results.

Authors' reply

Yes, the bulk of our seal testing on static force seals has been carried out having the bolts and bolting loads designed to strain under the pressure force and make an extrusion gap open up cyclically as the pressure is cycled. This method is not easy to control repeatably on small diameter seals and relatively low pressures since the low forces call for very small bolts and light bolting preloads, and for the tests reported in the paper we got more consistent perform- ance by having the bolts very rigid and the extrusion gaps machined in to the blocks. However, the pressure was cycled and we got results very similar to those which would be obtainable by allowing the blocks to "pant" due to bolt strain.

PAPER F2

Numerical and experimental stress–strain analysis on rubber–like seals in large elastic deformations under unilateral contact.

G. Medri, P.G. Molari and A. Strozzi

Bologna University, Italy.

J.P. O'DONOGHUE, NORTHERN IRELAND POLYTECHNIC, U.K.

I congratulate you on the elegance of your elasticity solution. If this technique is to be extended into dynamic lubrication conditions it is essential that an accurate pressure profile is derived over the small zones at the ends of the contact, these extend only 2% of the total contact. Is the finite element method capable of giving such an accurate profile?

In 1966 Dr. Hooke and I published an approximate solution to the 0-ring distortion based on small distortion theory (Ref. 1 below). Are you able to indicate the limits of application of this simplified theory?

Reference

1. Hooke, C.J., Lines, D.J. and O'Donoghue, J.P.: "Elastohydrodynamic lubrication of 0-ring seals". Proc. I. Mech. E., **181**, Pt. 1, No. 9, pp. 205-223. (1966-1967).

Authors' reply

Thank you for your kind words. We think that, in general, it should be possible to predict, with the accuracy one needs, the pressure profile putting more and more nodes in the zone in which a very high degree of accuracy is required or employing more sophisticated elements. Dealing with this highly non-linear problem, however, we think that it is necessary to prove this statement checking the numerical results with experimental ones and this is what we are trying to do.

With the finite element method it is also possible to evaluate the interaction between the rubber stress field and the hydrodynamic sustentation.

Dealing with your second question, we think that the solution presented in Ref. 13 below, explained and developed in Refs. 14, 15, and 16, already reported by one of us in a survey Ref. 17, takes into account the non linearities due to large displacements, but it can be employed only with almost linear materials.

References

13. Hooke, C.J., Lines, D.J. and O'Donoghue, J.P.: "Elastohydrodynamic lubrication of 0-ring seals". Proc. I. Mech. E. **181**, Pt. 1, 9, pp. 205-210. (1966-1967).

14. Discussion of the above paper. Proc. I. Mech. E. **181**, Pt. 1, 9, pp. 210-223. (1966-1967).

15. Hooke, C.J. and O'Donoghue, J.P.: "Elastohydrodynamic lubrication of soft, highly deformed contacts". J. Mech. Engineering Science, **14**, 1, pp. 34-48. (1972).

16. Discussion of the above paper. J. Mech. Engineering Science, **14**, 6, pp. 426-428. (1972).

17. Favretti, G. and Molari, P.G.: "Solid 0-ring seals for static applications". Oleodinamica e Pneumatica - serie di art. dal, 10, (1971) to 8 (1972).

B.S. NAU, BHRA FLUID ENGINEERING, U.K.

Would you please elaborate on the nature of the relation mentioned on p. F2-20 between the hydrostatic part of the stresses and the volume change.

Did you find that the calculated results depend very strongly on Poisson's ratio - is an accurate value of this ratio essential?

Authors' reply

As written at the top of p. 22, we have assumed a law $W = C_1 (I_1 -3) + C_2 (I_2 -3) + C_3 (I_3 -1)^2$, the latter part containing the volume change effects. We have assumed $C_3 \triangleq 10^3 (C_1, C_2)$ as suggested by Lindley (3), the stress field at this level of C_3 does not depend on C_3, but this remains an assumption not based on experimental work. We have considered $I_3 = 1$ in the plane stress field (5) and the method worked properly because the equation of the hydrostatic pressure can be eliminated by substitution.

The Poisson ratio in finite elasticity is substituted by the invariant I_3.

T.C. CHIVERS, C.E.G.B., U.K.

It is very encouraging to see the analysis of 0-ring deformation being extended. However, Paper F3 argues that groove constraints are very important in dictating contact stress, and hence seal performance. In these situations symmetry is lost, and the solution more complex. Do you have plans to extend your analysis, and if so could you discuss them?

Authors' reply

The symmetry has been assumed not to extend the number of degrees of freedom but with the finite element method it is possible to study the whole section, assuming constrained nodes also on lateral wall. This is only a matter of computing time and memory required. If the Italian National Council of Researches (CNR) gives us sufficient financial support, we will extend the static analysis to cover the elastohydrodynamic process and to compare experimentally the results so obtained.

PAPER F3

The achievement of minimum leakage from elastomeric seals.

T.C. Chivers and R.P. Hunt

CEGB Berkeley Nuclear Laboratories, U.K.

P.G. MOLARI, UNIVERSITY OF BOLOGNA, ITALY.

1. How have you measured the peak stress? (With a Muller-like apparatus or with something else?).
2. What do you mean when you talk about seal elastic modulus?
3. What is the diameter of seals tested? (I ask this because seal compression is limited to 25%).

Authors' reply

1. We did not measure the peak stress. The stress distribution was derived from previously published papers as referenced in Section 4.
2. By seal elastic modulus we refer to the Young's Modulus which is often measured as the seal hardness.
3. The limit of 25% to seal compression was arbitrary and not a derived limit based on seal cross section, which was 5.3 mm (0.21 inches).

R. METCALFE, ATOMIC ENERGY OF CANADA, CANADA.

It was mentioned in the presentation that elastomeric seals have been damaged or destroyed by internal pressure after rapid depressurization. This has been noticed in permeable carbon seal rings. Do you have knowledge of the parameters and relationships relevant to this phenomenon?

Authors' reply

The phenomena of damage as a consequence of rapid depressurisation has been considered in some detail, and a preliminary internal report published (Mitchell, L.A. and Osgood, C., 1971, CEGB Report RD/B/N2048).

Essentially the process is governed by the rate of diffusion, the diffusional path length, the retained pressure and the rate of depressurisation. These factors define the induced tensile stresses which can produce damage. The application of such analysis to practical situations is complicated by the need to define the induced stress levels that can lead to cracking.

Problems arise since a very detailed analysis of the stress distribution within the seal is required, particularly the identification of areas already in tension, and more susceptible to damage.

G.J. THEEUWES, PHILIPS RESEARCH LABORATORIES, NETHERLANDS.

To lower the leakage of gases we normally use a grease between 0-ring and groove (in Stirling engines to seal He and H_2). Why don't you do it to lower the leakage?

Authors' reply

The application of grease or other fluid to infill surface roughness and hence reduce leakage is common. As far as our particular interests are concerned there are a number of reasons for not employing them. One is related to large plant and the ability to maintain clean conditions. Experience has shown that where such fluids are employed they can retain dust and other debris, and hence be counter productive. In other situations we have problems of material compatibility, and arguments can be avoided if these sealant materials are not employed. An advantage from our approach is that if fluids can be employed then our answers are pessimistic. However, it is necessary in such cases to guarantee cleanliness and compatibility.

G.W. HALLIDAY, GEORGE ANGUS & CO., U.K.

Could you please clarify Fig. 2 which shows shaft contact stress can be significantly below the applied pressure. I have always understood that under pressure rubber could be considered to act as a fluid and would transmit the sealed pressure to the contacting surface.

Authors' reply

The derivation of Fig. 2 is discussed in Section 4.1. Basically there are two components to the seal contact stress, one due to the initial seal compression and the other due to the retained pressure. If a seal is adequately restrained then the contact stress should increase with retained pressure as shown in Fig. 1. When the seal is not adequately restrained the retained pressure can lead to distortion of the seal which may reduce the stress. This reduction continues until gross leakage occurs or until the retained pressure is sufficient for the seal to slip at the sealing surfaces.

J.D. KIBBLE AND J.C. LEAHY, NATIONAL COAL BOARD, U.K.

You have produced a very interesting paper about some very real problems in seal design. In particular you raise the question as to how best to deal with blemishes in the sealing surfaces, either resulting from the machining process, or occurring accidentally in use. The surface finish value, Ra, of course masks the existence of localised defects which may be vital in their effect on the sealing performance.

Isolated pits in a sealing face may not impair sealing provided they do not extend outside the area of seal contact. Even if the pit is not completely filled by the seal, the pressure around the edges will be intensified by extrusion of the seal into it. Asperities have the opposite effect. Unfortunately 0-rings provide a limited contact width, with maximum pressure only at the centre of it, and the dimensions of that too must be taken into account when choosing the compression needed to provide sealing against a surface of given geometry.

Could you please give some more detail about the experiments with discreet scratches etched in glass. In particular, the direction and position of those relative to the seal contact area; the shape of their cross section, and details of leakage measurements if any.

Authors' reply

The points raised in this contribution are very pertinent to the application of seal research to plant. One of the biggest difficulties being the assessment of potential accidental damage post manufacture.

A problem of specific interest was one where leakage could be measured under initial conditions, but it was necessary to consider the consequences of a reduction in compression during service. It is impossible to define the consequences of a hypothetical scratch or defect on a surface, but an analysis was performed in order to bound the problem.

Parameters considered were the:-

(i)   defect geometry,

(ii)   elastomer and contact stress,

(iii)   fluid mechanics,

and an upper bound for the increase in leakage as compression reduced was established based on the initial losses and assembly. The boundary was established in terms of triangular cross section defects whose length was equal to the seal contact width. Seal deformation into the defect was based on equation (6) and contact stress from Ref. (5). Laminar gas flow assumptions were shown to be the most pessimistic and should be employed unless arguments in favour of turbulent flow could be substantiated.

Experimental substantiation of the analysis was then sought. Domination of the leakage by one flow path was ensured by introducing one radial scratch into a smooth perspex plate; the '0' ring was constrained in the opposing surface with an outer groove conforming to BSS 1806, and experiments conducted with internal pressurisation. The scratch geometry was varied by using various implements for their introduction, e.g. razor and knife baldes. It was not easy to measure scratch geometry accurately, but they presented a predominantly concave surface to the elastomer, hence only width and depth were determined using optical techniques.

The results are summarised in Fig. 1 below, which shows very good agreement between the theory and the experiments. Fuller details are contained in CEGB report RD/B/N3680 ("A basis for quality control of surface defects on flanges for elastomer seals", Chivers, T.C., Hunt, R.P. and Williams, M.E.).

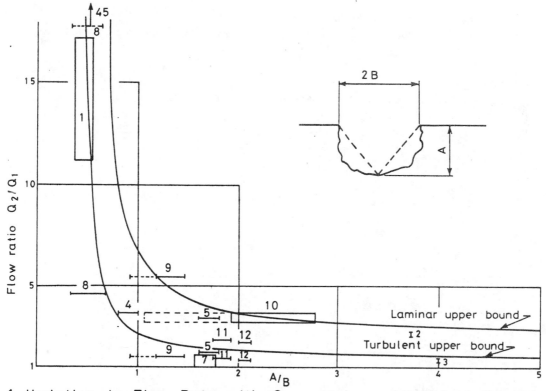

FIG. 1. Variation in Flow Rate with Compression and Defect Geometry. Initial Compression 17%. Final Compression 7%.

Discussion & Contributions

SESSION G : STATIC SEALS

Chairman: Mr. B.D. Halligan,
James Walker & Co. Ltd., U.K.

Papers:

G1   The testing of elastomers for channel seal plugs in a steam generating heavy water nuclear reactor.
J. Ward, N.P.C. (Risley) Ltd., and **K.A. Tomblin,** and M.L. Blakeston, AEE Winfrith, U.K.

G2   All-metal gasket for bolted flanged connectors.
H.J. Tuckmantel, Kempchen & Co. GmbH, Federal Republic of Germany.

G3   Scaling of gas leakage for static seals.
T.C. Chivers, and R.P. Hunt, CEGB Berkeley Nuclear Laboratories, U.K.

Notes:

1. The author in bold print presented the paper.

2. Errata on Paper G1 is provided overleaf.

ERRATA

PAPER G1
    Page G1-1

        In title of paper:-

        "REACTORS" should read: "REACTOR".

        The affiliation of J. Ward should read:-

        N.P.C. (Risley) Ltd., U.K.

Authors' replies follow the individual questions.

PAPER G1

The testing of elastomers for channel seal plugs in a steam generating heavy water nuclear reactor.

M.L. Blakeston and K.A. Tomblin

U.K.A.E.A., U.K.

and J. Ward

N.P.C. (Risley) Ltd., U.K.

R.P. HUNT, C.E.G.B., U.K.

1.   In Fig. 10 there is a suggested correlation between extrapolated "stand-off" measured at room temperature and seal leakage measured isothermally at 200°C. In light of the comments made by Mr. Aston in Paper F1, that leakage occurs earlier with temperature reduction, why should the results correlate?

2.   In operation the seals will be subjected to temperature reductions and hence failure before the point indicated in Fig. 10. Does this not mean that the "stand-off" extrapolation over estimates service life?

3.   Fig. 10 and Fig. 17 appear to suggest different relationships between the log of the elapsed time and either the "stand-off" or the compression set. If an Arrhenius relationship does not hold what is the justification for the data extrapolation in Fig. 10.

Authors' reply

1.   The results shown in Fig. 10 are not isothermal. Whilst the test rig ran for most of its life at a steady temperature it was cooled to atmospheric temperature for each measurement of stand-off. The values of stand-off so obtained were therefore in the cold state, and truly represented cold "start-up" conditions of the plant. It is at this stage of operation that leakage frequently occurs.

Our experience, although limited, indicates that extrapolation of curves based on cold stand-off figures, as in Fig. 10, provides a useful guide to the onset of leakage under start-up conditions.

2.   The wording in the box on Fig. 10 marking the onset of leakage is somewhat abbreviated: it refers to the first leakage of the seal which occurred when it failed to hold pressure on starting up at room temperature after the rig had been shut down. The normal rig operating sequence was first to pressurise at room temperature then heat up.

As explained above, the tests reported in Fig. 10 were designed to establish onset of leakage under start-up conditions rather than under undisturbed isothermal conditions.

3.   We claim no relationship between the parameters illustrated in Figs. 10 and 17. The Arrhenius plots in Fig. 17 are included for general interest. They represent an attempt to predict likely endurance of elastomer seals from early tests in air over limited periods. In practice the operating conditions of our seals were much more complex and it became obvious that only tests under true conditions would give reliable information. This led to the tests of which the results are shown in Fig. 10, which in our view much more closely resemble service conditions of the seals.

M.W. ASTON, LUCAS AEROSPACE, U.K.

1.   With reference to Fig. 12, when removing the plug how long was it before the stand-off was measured? Recovery will take place when the seals are not under constraint.
2.   Was there any stiction after long periods of time or at high temperatures?
3.   You are operating in water, yet your predictions are based on results in air. Your volume swells are positive in water and negative in air. Have you done any predictions based on water aging results?

Authors' reply

1.   This is a problem which affects most practical applications; we recognised the difficulty and tried to meet it by establishing standard conditions for measuring the rings in the laboratory apparatus. In this way the results for our tests are strictly comparable, but they are unlikely to be the same for other experimental conditions.

We arranged for the large seal plugs to cool down whilst maintaining hydraulic pressure, the plugs were then removed from their housings and stand-off of the rings measured within one hour. Quicker measurement was not easy to arrange, due to the bulky nature of the equipment and in any case practical checks on recovery suggested that it was relatively small over this interval of time.

2.   Measurements were made to establish the insertion and removal loads of seal plugs in the laboratory: no examples of significant stiction were observed.

In the reactor application measurements were attempted but any stiction is masked by the deadweight (about 400 Kg) of the fuel stringer which is directly attached to the seal plug.

3.   Each seal has to operate in two fluids, with air on one side and pressurised water on the other. In order to define limiting conditions initial tests were made in each fluid separately, and whilst the results were useful it was apparent that actual operational performance could only be assessed from sealing tests under full working conditions. The immersion tests in air and water were then used solely for comparative testing of different elastomers.

L. FRANGEUR, TRELLEBORG AB, SWEDEN.

1.   Choice of polymer: Have you found any difference between EPT and EPR in your application? I hope we are speaking about peroxide curing.
2.   Results from practice compared with the laboratory values: Don't you think your better results come from the fact you made a section of 6.99 instead of 3.9 mm? This gives you less contact with the oxygen in water and a better form factor.
3.   Deformation: Which deformation have you used - 25, 30 or 35%?

Authors' reply

1. Yes, we used peroxide cured EPT in our laboratory test programme and found it to be significantly better than EPR.
2. We agree that for true results the correct sized ring should be used, and in the later stages of the laboratory programme (Fig. 12) such rings were used. However, the moulding of large rings in a range of materials is expensive both in money and time, so we chose to conduct our earlier tests on materials which were more reasily available as small rings. The results from tests on small rings should not be taken to indicate a similar performance for larger rings: however, the small ring tests are valuable for comparison of materials.
3. In the immersion tests in pressurised water or air, an axial comparison of 15% was used.

PAPER G2

All-metal gasket for bolted flanged connectors.

H.J. Tuckmantel

Kempchen & Co. GmbH, Federal Republic of Germany

R.G. REES, I.I.R.S., IRELAND.

The use of two dissimilar metals in contact with the flanges, which may be of a third metal, in the presence of an electrolyte would normally be expected to cause corrosion of one of the metals.

Additionally the deep notches between the supporting ring and the flanges of an assembled interface appear to be the potential sources of further corrosion.

Are there then any limitations to the applications of H5 gaskets because of corrosion and are there any situations in which they should not be used?

Author's reply

The question of corrosion is always very complex. Comparing the H5 gasket with a silver-plated metallic 0-ring, one will encounter the above problem.

1. Flange, counterflange, 0-ring and coating often consist of three, or even four, different metals.
2. Also in the case of the metallic 0-ring, conditions favouring corrosion may arise within or near the contact area. As a rule, however, conditions prevailing at the H5 gasket are better than those prevailing at the metallic 0-ring with clip holes.

Hitherto no negative result of the application of H5 gaskets has come to notice. In this context it should not be left unsaid that, on account of its good thermal and mechanical resistance in addition to an almost unsurpassable gas tightness, the H5 gasket has so far been used predominantly for gas and vacuum applications.

Each intended application must be examined with a view to choosing the correct materials, but this rule applies to all gaskets. The combination of material No. 1.4541 and silver has proved successful in a wide range of application. Where there are chlorides or sulphides, material No. 1.4541 combined with aluminium can be used with good result.

J. ABBOTT, FLEXITALLIC GASKETS LTD., U.K.

The paper starts by listing four essential properties of a static seal, and then states that the only gasket which meets the requirements is a combination metallic gasket, i.e. the 'H5' gasket. How can this statement be justified when a) high pressure type 'R' ring joints to

API standard 6A and b) spiral-wound gaskets specifically designed for cycling pressure and temperature conditions have successfully served the chemical/petrochemical industry for many years?

Author's reply

It is correct that ring-joint gaskets and spiral-wound gaskets cover a wide field of application in the chemical and petrochemical industries. On the other hand, it must be said that there are cases where the H5 gasket offers more advantages. These must be seen especially in its extremely good stability (creep strength) together with extraordinarily low values of minimum design seating stress. Hence the minimum force per unit required for seating an H5 gasket with an aluminium sealing ring at 300°C is $\sigma_V = 32$ N/mm. Taking a gasket with one sealing ring on either side and an effective gasket width of 9.6 mm, this results in 307 N per millimetre circumference. With $\sigma_{\vartheta} = 420$ N/mm, the same gasket can withstand a maximum load per mm circumference of 4032 N. For H5 gaskets equipped with two sealing rings on either side the admissible load per mm circumference is almost double the afore-mentioned value.

To a user of high-pressure installations, for instance, it is of interest that FF and RF flanges can be used because this will considerably facilitate the assembly and, if necessary, disassembly of a joint. To him it is important that he can use an all-metal gasket on RF flanges.

P.G. MOLARI, UNIVERSITY OF BOLOGNA, ITALY.

1. What is the roughness of flanges and the particular texture?
2. What happens with thermal shocks of many Celsius degrees due to the very limited radial elasticity? Is there not a grinding effect?
3. Have you seen failure of the supporting ring with geometry of Fig. A due to cycling thermal loading?

Author's reply

1. Flange roughness has a direct bearing on the leak rate. The recommended maximum peak-to-valley height of 6.3 μm results in a leak rate of:-

$$L < 10^{-8} \; \frac{\text{mbar . cm}^3}{\text{s}} \, .$$

   Where demands were less severe, H5 gaskets were used with flanges of a peak-to-valley height of $R_F = 100 \; \mu$m. Especially in the case of important depths of roughness the scratches should run in peripheral direction.
2. Radial differential expansion as it may result from thermal shocks is detrimental in all sealing systems. It is often possible to avoid these radial movements by constructional measures. However, where these are an unavoidable nuisance, the principle applies that the finer the surface finish the better the sealing effect and the less wear or abrasion.
3. Failure of the supporting ring has not yet been observed.

Scaling of gas leakage for static seals.

T.C. Chivers and R.P. Hunt

CEGB Berkeley Nuclear Laboratories, U.K.

L. FRANGEUR, TRELLEBORG AB, SWEDEN.

Re. Fig. 7, the fluorocarbon in the figure must be older types like Viton A, Viton B or Technoflon T.

The new types like Viton E60C, Viton B910 or Technoflon FOR have a much better life time counted as compression set or compression relaxation.

Even better are the newest types like Viton GH or GF.

D. HUHN, GUSTAV HUHN AB, SWEDEN.

In Fig. 7 of the paper the expression "life-time" is used which is not explained anywhere else in the paper.

This type of information of life time for elastomeric materials in correlation to the temperature is essential for both manufacturers and users of any type of seals using rubber as a material in one singular or a multitude of components.

Could you specify the exact parameters behind the expression "life-time".

Authors' reply to L. Frangeur and D. Huhn.

Life time for any material is a function of the required duty. For seals the life will be dictated by acceptable leakage and such things as thermal movements. However, in order to obtain a design guide on life expectancy we adopted a 90 to 100% compression set criterion as defining end of life. We then abstracted appropriate data from the literature, as well as conducting our own experiments at the higher temperatures. At those temperatures where life is tens of minutes visual observation of deterioration was adopted to define failure. This is, of course, very arbitrary but is of relevance only to the prediction of behaviour in the presence of thermal transients.

In abstracting data from the literature no attempt was made to distinguish between individual specific polymer types because of the pausity of data. To establish one data point in Fig. 7 requires several values of compression set at the appropriate temperature, and preferably with some data at or near the failure point.

In view of the mixed origin of the data it is surprising that consistent trends are discernable for generic polymers, and it is also appreciated that the longer term data cannot include the more recently introduced mixes. However, currently there is insufficient information available to construct curves such as Fig. 7 for these new materials.

We would not suggest that universally definitive life time-temperature curves can be produced. Curves such as those shown in Fig. 7 should only be considered as a design aid to assist material selection for further consideration. In those situations where life assessment is critical, detailed failure analysis in terms of leakage and movement must be made. This aspect is considered in more detail in a report currently being prepared at Berkeley Nuclear Laboratories.

Discussion & Contributions

SESSION H : MECHANICAL FACE SEALS - III

Chairman: Mr. G.W. Halliday,
George Angus & Co. Ltd., U.K.

Papers:

H1   A theory for mechanical seal face thermodynamics.
P.C. Barnard and **R.S.L. Weir**, Esso Petroleum Co. Ltd., U.K.

H2   Study of electrodrill butt-end mechanical seals at high pressure differentials.
E.N. Griskin, Ministry of Electrical Engineering Industries of the U.S.S.R.,
U.S.S.R.

H3   Mechanical face seal for high pressure dredgepump.
E. Mudde and T. Visser, Mineral Technological Institute, The Netherlands.

Notes:

1.   The author in bold print presented the paper.

2.   Paper H2 was presented by R.A. Grimston, Flexibox Ltd., U.K.

Authors' replies follow the individual questions.

PAPER H1

A theory for mechanical seal face thermodynamics.

P.C. Barnard and R.S.L. Weir.

Esso Petroleum Co. Ltd., U.K.

Authors' General Comments

We should like to thank all the delegates who raised questions and comments on our paper and would like to take this opportunity to clarify four aspects which do not seem to have been fully understood.

1. **Leakage:** In real life applications of mechanical seals, most pumped streams contain dirt of an abrasive nature, this is particularly true of the oil refining industry. If seals are allowed to leak as predicted by the current flow base theories then the dirt in the product being sealed will inevitably migrate to the seal faces, be ingested into them and severely shorten their service life. The fact that seals commonly run for 2-4 years and some much longer would suggest that this dirt ingress does not occur. Our theory of fine face separation, a gas band, and ostensibly no leakage is consistent with these seal lives.

2. **Seal Face Separations:** Seal face separations in the order of 30 mm quoted in the paper was given to indicate the possible order of magnitude face separations that would be necessary to hold back the high pressure gas region by surface tension alone. It is not claimed by the authors that surface tension is the sole means of holding back the pressure nor that the fine face separations exist across the entire face width. Indeed it is anticipated that at these face separations, intermolecular forces play an important part especially with the long chain hydrocarbons found in refinery applications. This may well help to explain why water and light hydrocarbons with their short molecular chain lengths are more difficult to seal because they require face separation of possibly only 10 mm which is less than the surface texture of the faces.

3. **Face Distortion:** During operation seal faces are subjected to thermal, mechanical and pressure distortions. It is considered by the authors that all these effects take place to a greater or lesser degree in all seals and that once the seal has "bedded" in, they will have little further effect on the face performance unless there is a change of sealed pressure

effecting the attitude of one face to the other, the resulting mechanical movement has to be accommodated by the faces "re-bedding" in before satisfactory seal performance can once again be established. Leakage induced during these changes can often inflict damage to the faces that cannot be healed.

4. **Seal Face Surface Waviness:** One widely popular theory of face conditions relies or surface waves in the faces to carry the face loading in a manner similar to that of a taper land bearing. Two problems have given concern to the authors.

   (i) It relies on surface separations at a minimum of 2 nm when two adjacent wave crests are inline to 4×10^3 nm when the two valleys are adjacent, as can be appreciated this would give rise to large leakage rates and short seal life as noted in (1).

   (ii) Any seal that had worn itself into a wavy face would either leak on shutdown or, if the pressures were sufficient to flatten out the faces, cause very high face strains such that the face would suffer cracking.

M.T. THEW, SOUTHAMPTON UNIVERSITY, U.K.

Conclusions 1-4 do not seem contentious, but the fifth conclusion does bring in some new - or partially new - ideas, though I am not convinced that the three bands are universal. The ideas and evidence put forward appear to justify attempts to measure some of the parameters and thus to support or refute the conjectural distributions of temperature, pressure and face spacing.

With the implied phase change and the temperature distribution shown in Fig. 2 could you please sketch the variation in fluid viscosity across the faces radially.

Hence, with the aid of Fig. 5, could you please sketch the variation in the rate of heat generation across the faces. Then, having sketched in the heat source could you sketch in the heat flow paths for the whole of the stationary and rotating rings. I have suggested the above additions because it is difficult to see how the effects interact.

With the very large pressures postulated for the centre track region e.g. 218 bar (g), for water, it is difficult to see what retains any appreciable amounts of fluid between the faces there, yet if there was boundary lubrication would there not be wear?

The inter-face spacing distances suggested in the paper do not seem compatible with some "one or two light bands flatness" for first installation. Assuming newly installed seals do not leak do they operate in a different sealing regime until attaining profiles like Fig. 4 (hence Fig. 5)?

Authors' reply

We thank you for your comments and questions as they give the opportunity to further clarify some of the concepts.

It is of primary importance to appreciate that the face separation is not constant across the face width.

a. It is anticipated that the fluid viscosity will vary as shown in Fig. 1. The inside and outside liquid viscosities will depend upon the surrounding temperatures which will be a function of the heat being extracted from the faces by the fluid. This viscosity change will probably be in the order of 10:1.

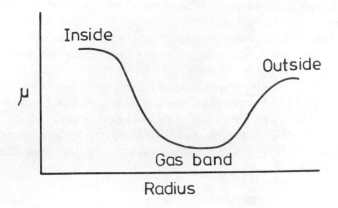

b. Heat generation per unit volume is proportional to viscosity and inversely proportional to seal face separation.

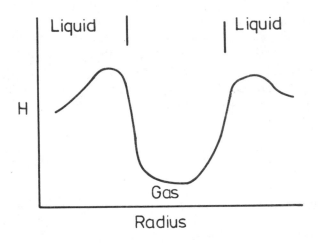

c. Heat flow is basically in 2 directions:-
From the liquid bands to the gas band.
From all 3 bands into the faces which are cooler than the interface film.
There is a net excess of heat production in the liquid bands over that conducted into the faces in these regions. This excess maintains the gas band temperature where there is insufficient viscosity to generate enough heat to match losses to the faces.

Certainly at temperatures as high as those which you postulate for the interference there would be thermal distortion producing 'coning', probably the sealed pressure would also cause 'coning'. The reference to a face separation of 0.00003 mm, necessary for surface tension forces to retain typical sealed pressures, seems very unrealistic when one considers that initially a commercial seal is likely to have circumferential waviness of amplitude about 0.00050 mm and a radial taper of nearly as much. Furthermore, any hydrodynamic pressures in a film as thin as that hypothesised could be quite enormous and could certainly not be ignored.

We ourselves have seen used seals with three zones (see Fig. below) but measurements showed considerable circumferential surface waviness, seven light bands of the example illustrated even though both surfaces were initially flat to a fraction of a light band.

Support for the view that angular distortion was present is provided by your slide showing a radial Talysurf trace which you showed in your presentation. Angular rotation of the upper surface provided an excellent fit to the wear tracks of the lower surface. (Fig. 4 shows another similar but less obvious example).

Finally, Fig. 7 shows a pressure profile which surely cannot be correct. If the pressure were truly higher at mid-face and lower inside and outside, and axisymmetric as your theory requires, then there must be radial flow tending to eliminate the pressure peak.

Could you please elaborate on why they feel that the Orcutt model is inadequate?

1200 rpm, 100 lb/in², Tellus 27 MY3B (Carbon) – Mechanite. Unit Load 130 lb/in²

Fig. 1   Carbon Face Without Optical Flat: Note Metallic Deposit

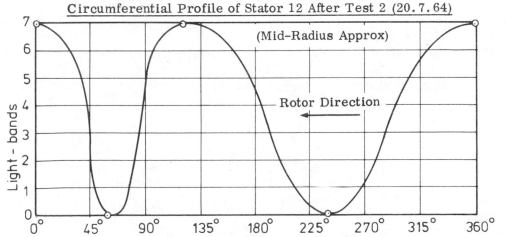

Fig. 2   Double Mechanical Seal: Test 2 (20.7.64) – After Running (5 hr), Stator 12

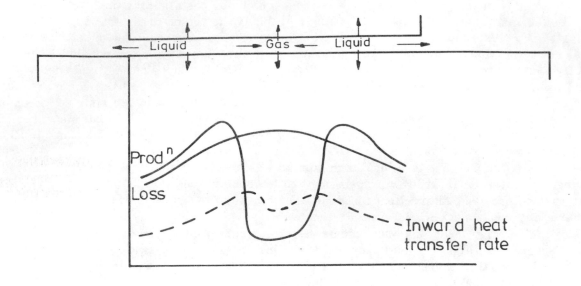

d.   To answer your last point, one needs to appreciate that seal faces need to be lapped to one to two light bands flatness to hold liquid in the static state. With some of the lighter hydrocarbons, a flatness figure of better than one light band is necessary to achieve a non leaking seal before startup. Once a seal starts to move, the faces will distort as outlined in the paper. The face as a whole is no longer flat but the proximity of the faces within one concentric band can be less than that achieved by the original faces. It is an interesting observation that seals that leak prior to start up rarely bed themselves in on running. We suspect this is because the leakage across the faces keeps them cool and prevents the build up of face temperature and hence the necessary face distortions.

M. LAGERQUIST, SANDVIK AB, SWEDEN.

Being a representative for a tungsten carbide manufacturer I must comment on the heat conductivity for tungsten carbide which is in the range 37-70 W/m^oC. What sort of carbides are you referring to, as straight cemented carbides have a heat conductivity in the range 90 - 120 W/m^oC.

Authors' reply

We have gathered our data from a number of sources including U.K. seal vendors and B.H.R.A. reports. Individual grades of Tungsten Carbide may well deviate in their properties from the typical data listed but the values quoted should not have a catastrophic effect on seal operation.

B.S. NAU, BHRA FLUID ENGINEERING, U.K.

You do not mention other work on the sealing of volatile liquids but that of Orcutt (1969 Int. Conf. on Fluid Sealing) and of Hughes et al (Trans ASME, 100, 1, pp. 74-80) is surely relevant. In the model explored by the above, a liquid film penetrates part of the way across the contact and then vapourises. You postulate a second liquid zone in which the vapour condenses near the low pressure edge but there seems to be no reason why the vapour should condense. You have been led to this assumption from the observation of three different wear zones observed on successful face seals. However, there could be a different explanation of these, for instance there could be different amounts of angular distortion initially and in the steady state.

R.T. ROWLES, BHRA FLUID ENGINEERING, U.K.

You have presented an interesting paper which, because it is based on practical observations, requires careful study. Clearly, the observations made can be compared with the observations of Orcutt, Ref. 1 below) and the calculations of Hughes et al (Ref. 2). Have you considered that:-

1.   Since you have not measured evaporation leakage, this leakage may be high enough to allow boiling of the fluid at a much lower temperature and pressure than the critical point.

2.   At the critical pressures quoted in the paper a considerable strain would be expected leading to local deformations well in excess of the film thickness quoted as operating values. This would be particularly true for the carbon rings mentioned.

It is to be hoped that you will follow through with your theory to include other observed effects such as hydrodynamics and coning that occur in other successful seals.

References

1. Orcutt, F.K.: "An investigation of the operation and failure of mechanical face seals". 4th International Conference on Fluid Sealing, Philadelphia USA, BHRA Fluid Engineering, Paper 22, pp. 205-217, (6-9 May, 1969).
2. Hughes, W.F. et al: "Phase change in liquid face seals". J. Lubrication Technology, 100, 1, pp. 74-80, (January, 1978).

Authors' reply to B.S. Nau and R.T. Rowles

We thank you for your comments and by way of answer would say that the work of Orcutt and Hughes is only half the story. If indeed the seals in many hydrocarbon services did operate with boiling taking place between the faces then quite large leakages should be observed and atmospheric side fouling found. The evidence is that on many hydrocarbon seals, that have slight leakage, carbon deposits are found on the atmospheric side indicating that the product that has leaked across the faces has been subjected to temperatures in excess of normal boiling temperatures. Often this product has thermally cracked - the thermal dissociation of hydro-carbons into lighter hydrocarbon fractions together with the deposition of carbon. The thermal cracking debris indicates that the seal face temperatures are probably 150-200°C above boiling point. On those seals that are successful there are no atmospheric side deposits let alone thermal cracking debris indicating no leakage and face temperatures above boiling yet below cracking. Hughes theory does not allow for temperatures as high as these.

Regarding the pressures involved and the possibility of face deformation the authors agree that this is possible but as the pressure is in circumferentially symetric rings the faces are not subjected to cyclic strains.

R. METCALFE, ATOMIC ENERGY OF CANADA, CANADA.

You have clearly been collecting good data on performance of refinery seals, but your paper would surely have been more useful if the facts had been presented in accurate details, with interpretation being left up to the reader. A particular concern is that many statements, including conclusion points 3, 4 and 5, may be read as generalizations rather than as possibly being relevant to one type of seal in one service. Such statements do not further our wider understanding of seal performance.

Some of the major questions and details that could be clarified are as follows:-

(i)   What are the pressures being sealed (e.g. for each result in Fig. 6)?

(ii)   What types of seals are being used - balanced or unbalanced?

(iii)   What is the sealed fluid for each result, Fig. 6?

(iv)   How clearly can track width be seen and measured, particularly width down to 0.1 mm?

(v)   Why do the inner and outer limits of the "wear track" not show on the profile trace (Fig. 4) of the hard face? Is the seal mounted eccentrically or is this profile not typical?

(vi)   Why has axial expansion of seal face material been discussed and sectional twist ignored? The effects on face deflection are approximately in the ratio of seal ring axial thickness to diameter, i.e. sectional twist several times more significant?

(vii)   How is a face separation as low as 0.00003 mm possible either with commercially "flat"
    components or with the excessive viscous shear that such a liquid film would generate?
        I hope the authors will not be disheartened by the discussion their paper has generated.
An orderly presentation of the data, some statistical interpretation, with empirical methods for
making specific seals work in specific service conditions, would be a valuable future contribution
to sealing technology.

Authors' reply

        Your comments suggest that you have failed to understand the nature of how we anticipate
seal faces to work, also the intent in writing the paper. The conclusion statements 3, 4 and 5
are stated because they lead to understanding the thermodynamics and mechanics of the face
operation. Statement 3 for example precludes any use of flow equation, to determine interface
pressures and temperatures, which is a major change from current thinking, Statement 4 indicates
that the interfilm condition must be not only stable at one set of conditions but able to absorb
variations in load and temperature, without wear, as seals do in the real world. Statement 5,
if accepted, says that previous theories of face waves, asperities, swash etc., previously
considered essential for correct face operation are all signs of bad seals not good ones. As to
just presenting facts and letting the reader draw his own conclusions, we find this at variance
with normal paper presentation practice and also that the main purpose of this paper was to
get people thinking in another direction!
        To answer your questions in detail:-
  (i)   The pressures range from atmospheric up to 20 bar g.
 (ii)   Both balanced and unbalanced.
(iii)   From crude oil to LPG and water.
 (iv)   When the seals are first removed from the pumps relatively easily, and they dull off with
    time after a week or so not so easily.
        Measurements on narrow tracks were made using a measuring microscope.
  (v)   At the outer edges of the faces there is erosive type wear which will degrade the softer
    face. The hard face often shows little or no wear at this point. On most of the machines
    under consideration the shafts were mounted in ball and roller bearings and as such the
    vast eccentricities that you consider necessary to make a seal work are not present.
 (vi)   Twist, coning and other face rotations were discussed during the presentation and are
    not ignored. These have been commented on in the general comment 3 at the begining of these
    replies.
(vii)   This question is answered in our reply to Mr. Thew's questions.
        With regard to your hopes that we will not be too disheartened by the discussion that the
paper generated, the converse is true. We have been much encouraged by the comments made
for they indicate that many of the people in the industry who have worked with single fluids and
seals that leaked, employing mechanical type solutions to their problems, are at last beginning
to think thermodynamically. Until the physics of mechanical seal face and inter film fluids
are understood there is little point in merely adding to the already long list of service examples.

J.P. O'DONOGHUE, NORTHERN IRELAND POLYTECHNIC, U.K.

        You claim to present a concept of lubrication unhindered by extensive calculations in the
hope that this will promote research to quantify your theory. Quantification is difficult in the
absence of any operating details. Fig. 6 shows 29 results from field tests. The value of such
information in terms of full information on the conditions of operation and the state of the seals
on removal is beyond measure in that it allows a careful consideration of the results against the
various theories. Sadly these details are missing, and instead we have a theory which assumes
a film thickness of 30×10^{-9} m, and a positive pressure gradient opposing fluid entry into the
gap. The energy, assuming a viscosity of 0.1 cp, is of the order of 1-10 KW, to be dissipated
by convection from the sides of the seal. It is very possible the oil will vapourise, and having
vapourised the shear will be reduced. Even without the surface tension effect leakage will be
negligible, the assumption of vapourised oil at atmospheric pressure leaving the seal gap at

sonic velocity still results in a leakage rate of the order of 1 cc per month!

To get this contact in perspective, if on Fig. 4 one measures one thirtieth of the 1 m thickness indicated at the side of the Talysurf traces, this is the separating film. The interstitial flow paths around the asperities of the steel surface are an order of magnitude greater than this value, consequently the analogy to the small bore tube filled with bitumen is unrealistic.

Had you calculated the necessary convective heat transfer coefficients necessary to cool the side faces of the seal, and if you had evaluated the friction energy due to fluid shear over even a small percentage of the seal face, then it would have shown that this theory is somewhat removed from practical reality.

Is it possible for you to give the following details of your successful seal application: seal type and material, speed, sealed fluid, sealed pressure, operating life and conditions of the faces on removal.

Authors' reply

As has been commented on earlier, you have made the assumption that the liquid film is or uniform thickness and viscosity across the entire face width and as such computes high heat generation figures, it is perhaps worthwhile correcting some of these misconceptions on fluid properties and scale factors.

The liquid surrounding a typical hydrocarbon seal would be around 1 to 2 cp; on vapourising to a mixed phase state its viscosity will drop to around 0.015 cp dropping to 0.01 cp when reaching the gas state at pressure. If one now computes these viscosities with the track widths given in the paper for a 3000 rpm 65 mm seal other results will give a film power loss of 35 watts which is very close to published data by seal vendors for such a seal on test. The heat flux and thermal gradients are well within the realms of possibility when one considers churning heat input by the rest of the seal components.

In answer to the second point, the interstitial flow path height is the same order of magnitude as the molecular diameter for many hydrocarbons. By only considering water, one is apt to overlook this, and we feel sure that it is molecular lengths which hold the key to why some fluids are so much easier to seal than others.

PAPER H3

Mechanical face seal for high pressure dredge pump.

E. Mudde and T. Visser

Mineral Technological Institute, The Netherlands.

J. VAN DEN BERG, MINISTRY OF DEFENCE, NETHERLANDS.

1. What was the reason for the Mineral Technological Institute to invite two laboratories for research on one and the same subject.
2. In the paper an increase of leakage of the seal installed at the "Mubarak" is mentioned after a period of 4000 running hours. This increase is from 1 l/h to 90 l/hr.
    Can you explain the reason for this increase? Did the increase occur gradually and if so could you give an idea of the amount of leakage increase versus time? What did the MTI find when they opened the seals? What is in their opinion the reason of this leakage-increase?
3. Why did you require a one-piece face ring for your seal? Had this something to do with assumed wear of split seals?
4. What are the experiences of the Mineral Technological Institute with the extrapolation of test results of small shafts (70 mm) to larger shafts (440 mm)?

Authors' reply

1. We have invited two different laboratories for the reason that the tests are different.
2. During the first 4000 running hours the seal was running in good order. Then the leakage started, but it was due to the job where the dredger was involved - it was impossible to stop the dredger earlier than 90 h from the start of the leakage.
    Through the fact that the leakage period was too long, the seal was damaged when opened. Normally such a seal can be reconditioned.
3. The assumption is right. A split in a seal surface is dangerous.
4. We have shown that increasing of the seal diameter needs a new seal design.

A.C. PIJCKE, NETHERLANDS MARITIME INSTITUTE, THE NETHERLANDS.

1. Could you give data on the proportion of the dredgepump price to the price of the applied mechanical seal.
2. Could you give your experiences regarding the technical co-operation between the dredge-pump manufacturers and the seal manufacturers. In this respect our experiences were very disappointing.

Authors' reply

1. Proportional to the price of a pump, the price of the seal is 10 : 1 - 5 : 1.
2. Our experiences with the seal manufacturers are also disappointing. There are however some better sealmakers.

Authors' reply to comments by A. Cameron-Johnson.
(for comments see Session A, page Z11).

Our seal construction is not a cheap construction, but a good one. We do not intend to produce very small ones.

Discussion & Contributions

SESSION J : RECIPROCATING SEALS

Chairman: D. Huhn,
Gustav Huhn A B, Sweden.

Papers:

J1   High pressure seals in Stirling engines.
G.J.A.M. Theeuwes, Philips Research Laboratories, The Netherlands.

J2   Contact stress, friction and the lubricant film of hydraulic cylinder seals.
R.M. Austin, R.K. Flitney and B.S. Nau, BHRA Fluid Engineering, U.K.

J3   Hydraulic seals for mining equipment.
H. Hopp, Martin Merkel KG, Federal Republic of Germany, and M.J. Harwood, FTL Co. Ltd., U.K.

Notes:

1. The author in bold print presented the paper.

2. Errata on Paper J1 is provided overleaf.

ERRATA

PAPER J1

Page J1-1

Third line from end of summary:-

"... paper discussed..." should read:

"...paper discusses..."

Page J1-2

Ninth line from bottom of page:-

"... The seal is a rollstock..." should read:

"... The seal is a rollsock..."

Authors' replies follow the individual questions.

PAPER J1

High pressure seals in Stirling engines.

G.J.A.M. Theeuwes

Philips Research Laboratories, The Netherlands.

J.D. KIBBLE, NATIONAL COAL BOARD, U.K.

More information on the action of the rollsock pressure-regulating valve would be useful. Does the valve operate on each cycle of the piston? Would it do so if, as was stated, this sealing principle was used for a compressor?

Author's reply

The rollsock-pressure regulating valve does not work each cycle of the piston. In Ref. 5 you will find more information about the design of the rod and the housing to get a constant oil-volume below the rollsock, which means that the pressure is not fluctuating during one cycle.

So if you meet these requirements the valve is only acting for long term pressure fluctuations and if the same is done in a compressor design you can use this sealing principle too.

A.E. AXFORD, ICI LTD., AGRICULTURAL DIV., U.K.

Were the nitrided cast iron liners pretreated by the application of a fluon coating before being put into operation?

Author's reply

The nitrided cast iron liners were not pretreated by the use of a PTFE coating. After grinding to the mentioned roughness the liners are degreased and ready for use. A big wear was noticed during the run-in period which means, we assume, that a PTFE layer is formed on the liner.

R.K. FLITNEY, BHRA FLUID ENGINEERING, U.K.

You state in the text that the friction of the rollsock system is negligible. Have you run your friction test rig, without piston ring to check that the hysteresis losses of the rolling diaphragm is in fact negligible. I have been unable to find any reference to this aspect of the rollsock performance.

Author's reply

Some test runs were done without piston rings to measure the influence of the rollsock seal and the windage losses. It was found that these losses were negligible compared to the piston ring friction.

PAPER J2

Contact stress, friction and the lubricant film of hydraulic cylinder seals.

R.M. Austin, R.K. Flitney and B.S. Nau

BHRA Fluid Engineering, U.K.

A. CAMERON-JOHNSON, SAUER U.K. LTD., U.K.

The work reported is confined to purely rubber type seals. I would like to ask whether any research has been done on seals consisting of a plastic shoe with a rubber expander, and then, whether from these tests, or from the vast amount of experience residing in BHRA, if it can be said whether a plastic shoe will behave in the same way as the reported behaviour of rubber.

Authors' reply

You ask about the 'coaxial' type of seal as compared to 'all-rubber' seals. There are, of course, some important differences relating to the plastic shoe, usually PTFE. In the short term the important features are a relatively high elastic modulus and low resilience. These affect ability to produce hydrodynamic lubrication and ability to accommodate lateral movements or rod tolerances. In the longer term the important features are plasticity, affecting creep, and sensitivity to wear by abrasive particles. With PTFE, of course, the low friction when unlubricated can be an over-riding advantage for certain applications but our experience is that low leakage rates cannot be guaranteed during prolonged use and this must therefore be weighed up against the advantage of low friction in individual applications.

P.C. BARNARD, ESSO PETROLEUM CO., U.K.

Re. Fig. 7 - What do you antitipate is happening under the seal in the very thin film shown for the instroke.

Authors' reply

During the instrokes, about which you ask, the film is so thin that it is at the limit of resolution of our transducer, it appears to be at most a fraction of a micron, perhaps much less. In this situation the surface asperities of both seal and counterface are large compared with any film which might be present. It seems probable in this situation that boundary lubrication is an important factor, that is to say a very thin oil film approaching molecular dimensions adhering to one or both surfaces and so preventing actual rubber to metal contact. On the other hand one cannot rule out the possibility that a hydrodynamic film still exists over part at least of the seal interface.

PAPER J3

Hydraulic seals for mining equipment.

H. Hopp

Martin Merkel KG, Federal Republic of Germany

and M.J. Harwood

FTL Co. Ltd., U.K.

B.S. NAU, BHRA FLUID ENGINEERING, U.K.

The final slide shown showed wear in the central region of the seal contact. This is interesting in the light of observations we have made in the laboratory using optical inter-ferometric techniques. We found that when the seal stroke is less than two contact widths the central region of the contact remains unlubricated and subject to severe shear stresses. This is due to the mean film velocity being only half the relative velocity of the surfaces. Your observations appear to show practical confirmation of the same phenomenon.

I wonder if you could say what the speed of the small motion was in this application.

Authors' reply

The piston speed of the rope winding cylinders has not been measured exactly. However, it is known to us that the same is operating with longer strokes at relatively low speeds (below 0, 1 m/s) which are overlapped by irregular high-frequency oscillation of differing amplitudes. These features appear from the winding process and as well as the elongation of rope resulting from driving jerks under oscillation of the cages.

You have found that dry-running would be encountered with strokes that are shorter than twice the contact length. We, however, are not in a position to confirm this observation because corresponding test runs have not been made by us. Experiences gained with existing units, however, reveal that difficulties may arise with stroke lengths of short contact lengths as a consequence of dry-running. This is particularly true in regard to homogeneous elastomer seals. However, similar observations have also been made with fabric seals operating under heavy duty service conditions.

R.K. FLITNEY, BHRA FLUID ENGINEERING, U.K.

I should like to ask if you could give us details of the surface finish of the roof support cylinders discussed in part 1 of the paper.

Authors' reply

The surface finish on the sliding faces of the roof support cylinders is as follows:

sliding surface: Ra $\leqq$ 0,3 μm $\hat{=}$ Rt $\leqq$ 3 μm $\hat{=}$ CLA ~ 16 μ in

groove base: Ra $\leqq$ 1,8 μm $\hat{=}$ Rt $\leqq$ 10 μm $\hat{=}$ CLA ~ 75 μ in

groove side: Ra $\leqq$ 3,0 μm $\hat{=}$ Rt $\leqq$ 16 μm $\hat{=}$ CLA ~ 120 μ in.

J.D. KIBBLE, NATIONAL COAL BOARD, U.K.

The development of multi-component seals for hydraulic props has certainly been of great value. They have to operate on water with little or no additives and to seal on surfaces which may be dirty or corroded, but performance is very satisfactory. The fact that compact seals are available has made possible such designs as double-acting double-telescopic props.

We have more difficulty, however, with reciprocating pump seals for these systems. Have you any experience of solutions to this problem?

Authors' reply

The experiences gained in the German mining industry substantiate that with the application of suitable COMPACT-seals satisfactory sealing effect is achieved despite relatively heavy duty service conditions.

For the sealing of plungers it is very common to employ particularly designed seals made from specially coated fabric and extremely hard impregnation and as well as braided packings.

Excellent results have been achieved with braided packings during the past years.

Possible misalignments of the plunger can easily be bridged with the use of braided packings. Further, braided packings would be the most appropriate sealing devices to absorb a certain quantity of dirt available in the medium without resulting into excessive wear.

In view of their excellent lubricating properties PTFE-impregnated braided packings offer satisfactory results even in such of those instances where the medium to be pumped presents insufficient lubricating properties.

M.T. THEW, SOUTHAMPTON UNIVERSITY, U.K.

With phosphate-ester hydraulic fluids there have been some suggestions that rapid corrosion sometimes seen in regions of high velocity may be of an electro-chemical nature. Modification of the electrical conductivity of the fluid has been found to reduce the corrosion in some applications.

Have you seen any sign of corrosion in your applications where the seals have high relative velocities?

Do you have any experience of reducing corrosion/erosion with phosphate-ester fluids by modifying the electrical conductivity?

Authors' reply

Up to now we have gained no experiences from which we could gather that there would be a close interdependence between corrosion and the speed of the plunger. If no grave mistakes are being committed, corrosion will primarily occur in the gaps where moisture (also from the atmosphere) and air are being mixed. In several cases, e.g. with roof support cylinders, corrosion also takes place as a result of an improper design of the seals. In addition, corrosion may be encountered as well if metals are used that differ substantially so far as the electrochemical series is concerned and if all further conditions are existing to initiate an electrochemical corrosion.

M.W. ASTON, LUCAS AEROSPACE, U.K.

You mentioned the use of Viton seals to seal phoshate ester fluids. We use phosphate ester fluids in aerospace applications but find that Viton seals are unsuitable and have to use EPR seals. Could you comment.

Authors' reply

In the hydraulic field Viton-seals are being successfully used against non-imflammable fluids of the HFD-group. These fluids, mainly phosphate ester fluids, consist of a certain number of ingredients and are also known in connection with various hydrocarbides.

The fluids used in aerospace applications such as Skydrol attack Viton-seals to a greater extent than those indicated above. In view of the above explanations we propose the application of EPDM-seals.

General Contribution to Paper J3:
B.D. HALLIGAN, JAMES WALKER & CO. LTD., U.K.

In response to a request by the Chairman for information on developments in high pressure water/oil reciprocating pumps, I would like to comment as follows: In addition to PTFE yarn packings used in Germany, conventional moulded proofed fabric V-packings are employed widely in the U.K. It is accepted that nitride proofed fabric is generally unacceptable in average life terms but recent material developments have been evaluated by N.C.B. Bretby and the pump manufacturers which indicate life improvements of the order of 12 times.

Suppliers of quality seals to the Hydraulics Industry.

Over 2,700 'O' Rings of standard sizes listed.

Ranges include: U.K. inch to BS 1806. Metric to BS 4518. German, Swedish and French metric sizes. American sizes and materials to MIL specifications. A comprehensive range of standard and specialised rubber compounds to meet your requirements. Large stocks of standard sizes in standard rubbers.

'O' Rings – the most extensive range ● 'Fluorobon' High Temperature 'O' Rings ● Shaft Seals, including '2DR' Hydrodynamic ● Rectangular Section Close Tolerance Sealing Rings ● 'U' Rings and Distributor Seals ● Piston and Gland Seals ● Bonded Seals ● Rubber to Metal Bonded Components ● Custom mouldings ● Wipers – Light, Medium and Heavy Duty ● 'Seloc' Washers (locking and sealing) ● 'Selon' Washers (captive gasket) ● High Temperature Metal Seals ● P.T.F.E. Products ● "Dowprint" Hydraulic Circuitry ● Rubber/P.T.F.E. Co-Axial seals ● Plastic mouldings ●

Dowty Seals Limited,
Ashchurch, Glos. GL20 8JS
Tel: Tewkesbury (0684) 292441. Telex: 43163.

Dowty Corporation,
Staverton West, Sully Road, P.O. Box 5000,
Sterling, Virginia 22170, U.S.A.
Tel: (703) 450 5930. Telex: 824459.

Dowty France SARL
BP54 Aéroport du Bourget, 93350 Le Bourget, France
Tel: 284-04-14 and 284-60-70. Telex: 212 200 Dowty SF.

Dowty Equipment of Canada Limited,
574 Monarch Avenue, Ajax, Ontario LIS 2GB, Canada.
Tel: (416) 683 3100. Telex: 06981295.

Klöckner-Dowty GmbH
462 Castrop Rauxel 2, Wartburgstrasse 69,
Postfach 686, West Germany.
Tel: Castrop Rauxel (02305) 1641. Telex: 08 229 512.

Dowty Industrial Division Companies.

Z78

While this pump's been standing here its seals have traveled 13 times around the earth

The site was a petroleum plant; the pumps were multi-stage horizontal splitcase design with 3½" shafts; the fluid was water, suction pressure 50 psig; discharge pressure, 3,500 psig, 35°F; the speed was 5,500 fpm (6,000 rpm). And the results were astonishing.

At first report — 5,200 hours after they were set — eight Dura Seals had logged 325,189 miles at a rate of 11" per revolution. Or more than 13 circumferences of the earth. And done it non-stop. Without a second's down-time.

But that's not remarkable for a Dura Seal®: we build them to stand even greater speeds — up to 12,000 fpm. To take temperatures from cryogenic to 700°F and pressures up to 3,500 psig; And to withstand solutions of extreme causticity, volatility, and abrasiveness. Solutions like black liquor service, in sulphate pulp and paper plants, crude oil in pipeline systems, condensate and borated water in nuclear plants.

Dura Seals: wherever we set them, we set them to last. Contact Durametallic for an obligation-free analysis of your sealing problem, and ask for Catalog 520. Marketing Services, Durametallic Corporation, Kalamazoo, Michigan 49001 — U.S.A.

DURAMETALLIC®
For the Dura Difference

SUBJECT INDEX

NAME INDEX

LIST OF DELEGATES

Appel, C.	Koninklijke/Shell Lab.	Netherlands
Arnswald, W.	ABS-Pumps GmbH	F.R. Germany
Ash, P.	Durametallic	U.K.
Aston, M.W.	Lucas Aerospace	U.K.
Aubin	Le Joint Francais	France
Austin, R.M.	BHRA Fluid Engineering	U.K.
Axford, A.E.	I.C.I. Ltd. (Agric. Div.)	U.K.
Bailey, R.L.	Ferrox	U.K.
Ballard, D.	Shamban Europa (U.K.) Ltd.	U.K.
Banse, K..	Burgmann, Feodor	F.R. Germany
Barnard, P.C.	Esso Petroleum Co. Ltd.	U.K.
Barwell, F.W.	Swansea, University College,	U.K.
Bijlsma, M.	TBB Zandvoort BV	Netherlands
Blakeston, M.L.	U.K.A.E.A.	U.K.
Bollina, E.	Milano Politecnico	Italy
Brown, R.G.	U.S. Naval Ship R & D	U.S.A.
Cameron-Johnson, A.	Sauer U.K. Ltd.	U.K.
Cannings, J.A.	Cannings Seals	U.K.
Carrington, J.E.	Short Bros. Ltd.	U.K.
Chinnery, R.	Angus, Geo., & Co.	U.K.
Chivers, T.C.	C.E.G.B.	U.K.
Christ, A.	Escher Wyss	Switzerland
Cognet, M.	H.B.S.	France
Cook, P.E.J.	Simrit Ltd.	U.K.
Cooper, R.M.	Gen. Dynamics Elec. Boat Div.	U.S.A.
Coppendale, J.	True Investments Res. Labs.	U.K.
Course, H.	Bestobell Seals Ltd.	U.K.
Daly, J.D.	Borg Warner Mech. Seals	U.K.
De La Rey	Angus-Hawken (Pty) Ltd.	S. Africa
Deuring, H.	Goetze AG.	F.R. Germany
Doust, T.G.	Crane Packing Ltd.	U.K.
Edwards, S.J.	Pionier Laura BV	Netherlands
Ellis, R.G.H.	Ministry of Defence	U.K.
Enis, H.S.	Ministry of Defence	Israel
Flitney, R.K.	BHRA Fluid Engineering	U.K.
Frangeur, L.	Trelleborg AB	Sweden
Garner, P.J.	Hallite Holdings	U.K.
Gibson, F.J.	Walker, James, & Co. Ltd.	U.K.
Ginn, A.	IPC	U.S.A.
Goransson, L.	Atlas Copco Airpower NV	Belgium
Gough, P.H.H.	British Leyland Cars Ltd.	U.K.
Gove, K.B.	Associated Engineering Dev. Ltd.	U.K.
Greiner, H.F.	Sealol Inc.	U.S.A.
Grimston, R.A.	Flexibox Ltd.	U.K.
Guy, N.	BHRA Fluid Engineering	U.K.

Hafner, W.	Freudenburg, Carl	F.R. Germany
Halliday, G.W.	Angus, Geo., & Co.	U.K.
Halligan, B.D.	Walker, James, & Co. Ltd.	U.K.
Hanagarth, W.	Klein, Schanzlin & Becker AG	F.R. Germany
Harwood, M.J.	F.T.L. Co. Ltd.	U.K.
Heitel, K.	Stuttgart University	F.R. Germany
Hershey, L.E.	Durametallic Corp.	U.S.A.
Hickson, J.P.	B.N.F. Ltd.	U.K.
Hom, J.	Shamban, W.S., Europa A/S	Denmark
Hopp, H.	Martin Merkel KG	F.R. Germany
Hughes, A.	Bestobell Seals Ltd.	U.K.
Huhn, D.	Huhn, Gustav, AB	Sweden.
Huhn, P.	Pacific Wietz GmbH	F.R. Germany
Huish, P.K.	Compair Industrial Ltd.	U.K.
Hunt, R.P.	C.E.G.B.	U.K.
Johnson, C.M.	Flexibox Ltd.	U.K.
Johnson, N.E.	Sealol Inc.	U.S.A.
Johnston, D.E.	Angus, Geo., & Co.	U.K.
Jonsson, I.	Stefa Industri AB	Sweden
Joy, N.	Pioneer Weston Ltd.	U.K.
Kaneta, M.	Kyushu Inst. Technol.	Japan
Kennedy, P.	U.K.A.E.A.	U.K.
Kersting, A.J.R.	Fokker VFW BV	Netherlands
Kibble, J.D.	National Coal Board	U.K.
Labus, T.J.	Sealol Inc.	U.S.A.
Lagerquist, M.	Sandvik AB	Sweden
Laurenson, I.T.	Northern Ireland Polytechnic	U.K.
Lawrence, A.J.	Tungum Hydraulics	U.K.
Lee, R.	Angus, Geo., & Co. Ltd.	U.K.
Loewen, T.	Ontario Hydro	Canada
Luxford, G.	Crane Packing Ltd.	U.K.
McLaughlan, W.A.	Brit. Nuclear Fuels Ltd.	U.K.
Merlini, M.	Paulstra	France
Metcalfe, R.	Atomic Energy of Canada	Canada
Mewes, H.	Howaldtswerke Deutsche Werft	F.R. Germany
Meyer, W.M.	Sealol Inc.	U.S.A.
Miannay, C.R.	Exxon Co.	U.S.A.
Millar, T.R.	Crossley, Henry, (Packing)	U.K.
Molari, P.G.	Bologna University	Italy
Moodie, K.	Health & Safety Exec.	U.K.
Mudde, F.	Mineral Tech. Inst.	Netherlands
Mueller, H.K.	Stuttgart University	F.R. Germany
Nau, B.S.	BHRA Fluid Engineering	U.K.
Nelson, A.R.	Crane Packing Ltd.	U.K.
Neubauer, G.	Freudenburg, Carl	F.R. Germany
Nolan, D.R.	Angus, Geo., & Co.	U.K.
Nommensen, J.P.	Dutch State Mines	Netherlands
Nordin, D.	Flygt AB	Sweden
O'Donoghue, J.P.	Northern Ireland Polytechnic	U.K.
Ohtaki, M.	Nippon Oil Seal Industry Co. Ltd.	Japan
Organ, A.H.	Dowty Seals Ltd.	U.K.

Palmer, K.P.	Lucas Aerospace	U.K.
Parkinson, K.	Brit. Nuclear Fuels Ltd.	U.K.
Parr, N.L.	Consultant	U.K.
Pataille, G.	Procal	France
Paternoster, K.W.	Sealol Ltd.	U.K.
Pennock, T.A.F.	Neratoom BV	Netherlands
Phillips, J.	Flexibox Ltd.	U.K.
Phillips, J.J.	Bestobell Seals Ltd.	U.K.
Pierrot	Le Joint Francais	France
Pijcke, A.C.	Netherlands Maritime Inst.	Netherlands
Plumridge, J.N.	Sealol Ltd.	U.K.
Pretty, S.R.	CAV Ltd.	U.K.
Prevedini, B.	C.I.S.E.	Italy
Price, W.G.	Dowty Seals Ltd.	U.K.
Ravetta, R.	C.I.S.E.	Italy
Rayner, K.G.	I.C.I. Ltd. (Agric. Div.)	U.K.
Reddy, D.	BHRA Fluid Engineering	U.K.
Rees, R.G.	I.I.R.S.	Eire
Ridderskamp, F.	Freudenburg, Carl	F.R. Germany
Robinson, C.	C.E.G.B.	U.K.
Robinson, R.P.	Flexibox Ltd.	U.K.
Rogers, P.R.	Crane Packing Ltd.	U.K.
Rotter, A.	VEB Kosid-Kautasit-Werke	German D.R.
Rowles, R.T.	BHRA Fluid Engineering	U.K.
Ruppel, J.C.	Carborundum Co.	U.S.A.
Sargent, G.	Angus, Geo., & Co.	U.K.
Schmid, E.A.	Kupfer-Asbest-Co.	F.R. Germany
Schoepplein, W.	Burgmann, Feodor	F.R. Germany
Scott, P.A.J.	Univ. Metallic Packing Co.	U.K.
Spies, K.H.	Freudenburg, Carl	F.R. Germany
Staaf, F.	Swedish State Power Board	Sweden
Stephens, H.S.	BHRA Fluid Engineering	U.K.
Stephens, J.A.	Bestobell Seals Ltd.	U.K.
Straszewski, C.J.	Sealol Ltd.	U.K.
Strozzi, A.	Bologna University	Italy
Summers-Smith, D.	I.C.I. Ltd. (Agric. Div.)	U.K.
Summersgill, J.	Pioneer Weston Ltd.	U.K.
Swales, P.D.	Leeds University	U.K.
Taylor, J.	Crossley, Henry, (Packing)	U.K.
Taylor, P.	Ministry of Defence	U.K.
Telfer, A.	Ontario Hydro	Canada
Theeuwes, G.J.	Philips Res. Labs.	Netherlands
Thew, M.T.	Southampton University	U.K.
Tomblin, K.A.	U.K.A.E.A.	U.K.
Trepleton, P.G.	Aeroquip (U.K.) Ltd.	U.K.
Tribe, F.J.	Admiralty Marine Technol. Est.	U.K.
Trytek, J.J.	Crane Packing Co.	U.S.A.
Tueckmantel, H.J.	Kempchen & Co.	F.R. Germany
Tuinman, W.	Mineral Tech. Inst.	Netherlands
Tully, N.	Natal University	S. Africa
Turnbull, D.E.	Petroleum & Minerals Univ.	Saudi Arabia

Van Den Berg, J.	Min. Van Defensie	Netherlands
Verhey, L.A.	Neratoom, BV	Netherlands
Verheyden, L.	Euratom	Italy
Victor, K.H.	Pacific Wietz GmbH	F.R. Germany
Vignaud, M.	E.D.F.	France
Vries, P.	Royal Netherlands Naval College	Netherlands
Wakely, K.W.	Morganite Special Carbons	U.K.
Walde, A.	Durametallic Europe GmbH	F.R. Germany
Wallace, N.	Flexibox Ltd.	U.K.
Ward, J.	N.P.C. (Risley) Ltd.	U.K.
Waters, R.	Borg Warner Mech. Seals	U.K.
Weir, R.S.L.	Esso Petroleum Co. Ltd.	U.K.
White, C.N.	Angus-Hawken (Pty) Ltd.	S. Africa
Wiles, W.F.	Caulfield Inst. of Technology	Australia
Wing, B.	Aeroquip (U.K.) Ltd.	U.K.
Ziehe, G.	Kombinat Pumpen Und Verdichter	German D.R.